T0140315

Explainable Artificial Intelligence for Intelligent
Transportation Systems

Loveleen Gaur • Biswa Mohan Sahoo

Explainable Artificial Intelligence for Intelligent Transportation Systems

Ethics and Applications

 Springer

Loveleen Gaur ⓘD
Amity International Business School
Amity University
Noida, India

Biswa Mohan Sahoo ⓘD
School of Computing and IT
Manipal University Jaipur
Jaipur, India

ISBN 978-3-031-09646-4   ISBN 978-3-031-09644-0   (eBook)
https://doi.org/10.1007/978-3-031-09644-0

This Springer imprint is published by the registered company Springer Nature Switzerland AG
The registered company address is: Gewerbestrasse 11, 6330 Cham, Switzerland

This book is dedicated to our respective family, friends, students, and all those who have inspired us directly and indirectly.

Preface

Explainable artificial intelligence (AI) lies at the core of many activity sectors that have embraced new information technologies. Although the origins of AI can be traced back many decades, there is widespread agreement that intelligent systems with understanding, logic, and adaptation capabilities are of utmost importance today. AI approaches are reaching unparalleled levels of success while learning to solve exponentially complex computational challenges because of these capabilities, rendering them crucial for the future growth of human society. The complexity of AI-powered systems has recently improved to the point that their architecture and implementation involve virtually no human involvement. Transportation typically entails crucial "life-death" choices, and delegating crucial decisions to an AI algorithm without any explanation poses a serious threat. Hence, explainability and responsible AI is crucial in the context of intelligent transportation. In intelligence transportation system (ITS) implementations such as traffic management systems and autonomous driving applications, AI-based control mechanisms are gaining prominence. The book discusses and informs researchers on explainable intelligent transportation system. It also discusses prospective methods and techniques for enabling the interpretability of transportation systems. The book further focuses on ethical considerations apart from technical considerations.

Loveleen Gaur
Noida, India

Biswa Mohan Sahoo
Jaipur, India

Acknowledgments

Nothing is more important and urgent than giving thanks. Many individuals have provided their valuable suggestions and criticism, which helped publish the first edition. We appreciate the efforts of those individuals who provided a formal review of the book's first edition.

I am also grateful to Springer for showing confidence in our work. Without their help, the creation of this book would not have been possible.

Also, we would like to acknowledge with gratitude the support and love of our family and, finally, our gratitude to the divine power and HIS blessings.

Loveleen Gaur
Biswa Mohan Sahoo

Contents

List of Figures

List of Tables

About the Authors

Loveleen Gaur is currently working as a professor and program director of artificial intelligence and data analytics at Amity International Business School, Amity University, India. She has more than 20 years of teaching, research, and administrative experience internationally. She is heading MBA program in artificial intelligence and data analytics at Amity International Business School. She is supervising a number of PhD scholars and postgraduate students mainly in artificial intelligence and data analytics for business and management. Under her guidance, the AI/data analytics research cluster has published extensively in high impact factor journals and has established extensive research collaboration globally with several renowned professionals.

Prof. Gaur is a senior IEEE member and series editor with CRC and Wiley. She has *high indexed publications* in SCI/ABDC/WoS/Scopus and has several *patents/copyrights* on her account, and she has *edited/authored* more than *20* research *books* published by world-class publishers. She has excellent experience in *supervising and co-supervising postgraduate* students internationally. An ample number of PhD and master's students graduated under her supervision. She is an *external PhD/master's thesis examiner/evaluator* for several universities globally. Prof. Gaur has completed *internationally funded research grants* successfully. She has also served as *keynote speaker* for several international conferences, presented several *webinars* worldwide, and chaired international conference sessions. Prof. Gaur has significantly contributed to enhancing scientific understanding by participating in more than 300 scientific conferences, symposia, and seminars, by chairing technical sessions and delivering plenary and invited talks.

Biswa Mohan Sahoo is a senior member of IEEE. He has received his BTech and MTech degrees in computer science and engineering from Biju Patnaik University of Technology, Odisha, India, and PhD degree in computer science and engineering from the Indian Institute of Technology (IIT), Dhanbad, India. Dr. Sahoo is currently working as an assistant professor at Manipal University, Jaipur, India. He has 15 years of teaching experience in different universities in India. He has published more than 18 articles in prestigious international peer-reviewed journals and conferences on wireless sensor networks, swarm intelligence, and artificial intelligence. He is currently focusing on artificial intelligence approaches on sensor networks. His research areas of interest are wireless sensor network, IoT, and artificial intelligence.

Abbreviations

ACC	Adaptative Cruise Control
ACO	Ant Colony Optimization
ADAS	Advanced Driver Assistance System
AI	Artificial Intelligence
ANN	Artificial Neural Network
ANPR	Automatic Number Plate Recognition System
APB HD	Heavy-Duty Automated Parking Brake
APPs	Application Programming Interfaces
APTS	Advanced Public Transportation System
ATID	Advanced Traveler Information System
ATMS	Advanced Transportation Management System
ATPS	Advanced Transportation Pricing System
AVL	Automatic Vehicle Location
CA	Cellular Automata
CAI	Cornell Aeronautical Laboratory
CCC	Common Control Channel
CH	Cluster Head
CNN	Convolutional Neural Network
CRM	Customer Relationship Management
CVS	Co-operative Vehicle System
DAS	Driver Assistance Systems
DL	Deep Learning
DNN	Deep Neural Networks
DoS	Denial-of-Service
EBEESU	ElectriBio-Inspired Energy-Efficient Self-Organization Model
EER-SHO	Energy-Efficient Routing – Spotted Hyena Optimizer
ETS	Electronic Ticketing System
EUGDPR	European Union's General Data Protection Regulation
FMCW	Frequency Modulated Continuous Wave
GA	Genetic Algorithm
GHG	Greenhouse Gas

GPS	Global Positioning System
HACA	Hybrid Ant Colony Algorithm
HiTsec	Hierarchical Trust-based Secure Clustering
ICT	Information Communications Technology
IoT	Internet of Things
ITS	Intelligent Transportation Systems
KNN	K Neural Networks
LDWS	Lane Departure Warning System
LPR	License Plate Recognition
MDA	Mean Decrease Accuracy
MIE	Mean Increase Error
ML	Machine Learning
MLP	Multi-layer Perceptron classifier
NDP	Network Design Problem
NISMOD	National Infrastructure Systems Model
NN	Neural Networks
OBD	Onboard Diagnostics
PLADS	Parachute Low Altitude Delivery System
PNN	Probabilistic Neural Network
PSO	Particle Swarm Optimization
RL	Reinforcement Learning
RNN	Recurrent Neural Network
RSU	Roadside Units
SAE	Society of Automotive Engineers
SHAP	Shapley Additive Explanation
SHO	Spotted Hyena Optimizer
SP	Schedule Planning
STEL	Simplified Tree Ensemble Learner
SVC	Support Vector Clustering
SVM	Support Vector Machine
TP	Technological Partner
TTI	Traffic and Travel Information
UAV	Unmanned Aerial Vehicles
V2I	Vehicle-to-Infrastructure
V2V	Vehicle-to-Vehicle
VANET	Vehicular Ad hoc Networks
WSN	Wireless Sensor Networks
XAI	Explainable Artificial Intelligence

Chapter 1
Introduction to Explainable AI and Intelligent Transportation

1.1 Introduction

Explainable Artificial Intelligence (XAI) has become a central focus in many industries that have adopted new information technologies. Many people feel that intelligent robots with learning, reasoning, and adaptive abilities are vital today, even if the beginnings of AI can be traced back many decades. XAI methods may improve their performance and help them learn to handle increasingly challenging computational tasks, making them critical for human civilisation's continued progress. Recent advancements in XAI-powered systems have made their design and deployment possible to be carried out without any human intervention. There is a rising need to understand how such choices are made utilising AI approaches when they influence people's lives (such as in medical, legal, or defence) [1, 2].

Deep Neural Networks (DNNs) have become more prevalent since the early AI systems were easy to grasp. DNNs and other DL models are used in experiments because of their sizeable parametric range and rapid learning processes. A DNN is a complex black-box model because many layers and parameters contain transparency, the opposite of black box-ness, which is a desire to precisely know how it works [3, 4]. Transparency among various AI stakeholders is becoming more critical as black-box machine learning (ML) algorithms are being used to make essential predictions in urgent scenarios. The risk of executing and carrying out actions that are not backed by facts or do not adequately justify them [7]. For instance, physicians want a model to back up their diagnosis with significantly more information than a simple binary prediction [8]. It is critical to provide context. Self-driving cars find applications in various fields, from transportation to security to finance.

Explainability has been considered a critical aspect in the widespread acceptance of AI systems. An explanation for the automated decisions is crucial when used in application domains such as transportation and autonomous vehicles, medical diagnosis, insurance, and financial services for practical, social, and increasingly legal

L. Gaur, B. M. Sahoo, *Explainable Artificial Intelligence for Intelligent Transportation Systems*, https://doi.org/10.1007/978-3-031-09644-0_1

reasons for the deployment of intelligent systems. It's general knowledge that the European Union's General Data Protection Regulation (GDPR) gives consumers who are affected by automated judgments a "right to an explanation" or "meaningful information about the reasons involved." However, the motivations for providing explanatory skills to intelligent systems are not confined to user rights and technological adoption concerns. Designers and developers must also consider explainability to improve system resilience, allow diagnostics to minimise prejudice, unfairness, and discrimination, and raise user confidence in why and how choices are made. Thus, the capacity to explain why a specific decision is made has become a highly sought-after feature in intelligent systems. Users should be able to debug the system model with the help of explanations so that incorrect conclusions may be avoided and corrected. It helps users maintain and efficiently utilise the system model. Explanations may also serve as a teaching tool and assist in discovering and understanding new concepts in a particular application area. Explanations, in turn, are connected to trust and persuasion in that they should convey a sense of actionability and convince users that the system is making a good decision for them. Despite this, no one can agree on what makes a reasonable explanation or what constitutes an explanation. Many AI systems and areas have looked at its expressions. The explainability in AI has faded along with expert systems. Still, new achievements in deep learning technology for autonomous and human-in-the-loop systems, with applications in recommender systems and neural-symbolic reasoning, have brought them back into prominence [5, 6].

The chapter examines XAI literature from a historical viewpoint, including classic and emerging techniques. The chapter aims to present an overview and discuss how several conceptions of explainability (or explanation format) have been established and provide some instances in Fig. 1.1 [3].

A significant problem in several domains today, including cognitive science, computer science, psychology, philosophy, and mathematics, is determining the criteria for a good explanation, which is still an ongoing work in progress. By providing an in-depth examination of explanatory research in philosophy, psychology, and cognitive science, the book shows the connection between social scientific controversies and the explainability of artificial intelligence.

People understand why some events happened but not others because explanations are counterfactual. Explanations should limit themselves to one or two plausible explanations for a decision or proposal to avoid overwhelming the user with too much information. It is important to remember that explanations are a social conversation and interaction to impart knowledge, which implies that the explainer must be able to apply the explainer's conceptual model when explaining. While these three traits are essential for creating adequate explanations, according to Miller (2019), the various conceptions of explainability used in XAI have only lately begun to consider them [9]. Psychologists, for example, have investigated and characterised the characteristics of human-oriented explanations. They have stated that one should distinguish between multiple alternative aims for explainability and why and how human explanatory cognition imposes significant constraints on the design of XAI systems. According to the report, when employing

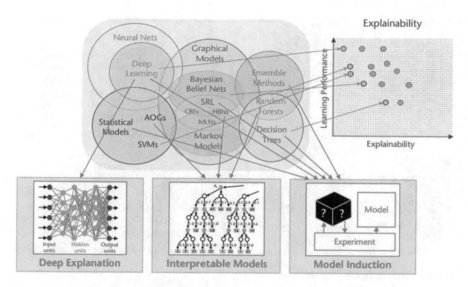

Fig. 1.1 Elaboration methods of Explainable AI

machine-generated explanations, it is vital to associate semantic information with an explanation (or its components) to maximise the efficiency of knowledge transmission to human users. According to Gilpin et al. [10], who investigated the relationship between specific qualities of produced explanations and the faithfulness of users' conceptual models, they discovered that completeness appears to be more critical than soundness. In the end, simplifying an explanation decreases users' confidence in the explanation. Until recently, most computer science research has focused on the mechanical aspects of constructing explanations. As well as knowledge-based systems, this applies to machine learning and recommender systems. The sorts of explanations these systems may produce – and hence their properties – are primarily determined by the style of reasoning used in the design, which can be symbolic, sub-symbolic, or hybrid [11].

Figure 1.2 illustrates that XAI can provide users with explanations that allow them to comprehend the system's general strengths and weaknesses, communicate how it will behave in future or different scenarios, and perhaps even allow users to correct the system's errors [3].

An inference mechanism (such as deduction, abduction, or analogical reasoning) and a knowledge base (usually recorded as a collection of products or symbolic rules) derive conclusions or explain why a hypothesis is correct. There are two types of explanations in these systems: descriptions related to the reasoning trail of the system or reports more directly tied to the narrative behind the system's decision-making process. The measures such as correctness, adaptability, and understandability are sought in explanations. For these explanations to be helpful to a wide range of people, they must be able to adapt to different user profiles. While professionals prefer more technical and extensive descriptions, average customers may choose a less precise but more straightforward explanation [12].

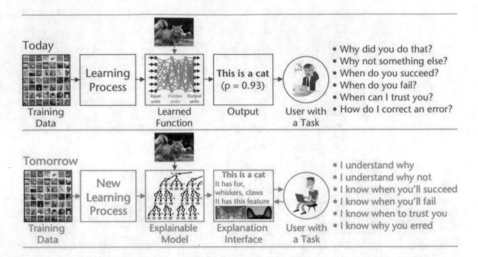

Fig. 1.2 The concept of Explainable AI

In general, sub-symbolic reasoning systems rely on machine learning models in which representations are scattered in most cases, and processing occurs in several parallel channels. Because of these qualities, the corresponding models are often left undocumented. Thus, in these systems, the explanations are generally presented as interpretable models that mimic or seek to duplicate the actual behaviour of their subjects. An interpretable model explains how decisions are made by providing local or global post-hoc explanations [15]. Metrics like accuracy and fidelity are often used to evaluate interpretable models. Holzinger et al. [21] developed additional metrics aimed at the causality of explanations; that determine how well an interpretable model can retain competitive levels of accuracy compared to the original black-box model and how well it can successfully simulate a black-box prediction. Efficacy, efficiency, satisfaction with causal understanding, and user transparency are criteria used to evaluate how an explanation meets a specified level in each use context. Recently, counterfactual explanations have been studied as an additional class of explanation strategies for black-box models, particularly determining alternative decisions or predictions.

Among the many types of sub-symbolic systems, recommender systems fall. There is no consensus on what makes a good explanation in the study on recommender systems. Any suggested resolution may be used for different goals and have other consequences on the people who make the decisions. A personalised description, for example, may help a consumer find an item more quickly. To begin producing an explanation, it is typically first to identify its purpose. There may be an incentive in persuasion from stakeholders since it increases the chance of acceptance. Because users are more likely to return to and reuse systems they feel confident in, credibility is essential in an explanation. In addition, clear, concise, and enjoyable explanations help users make fast, accurate decisions and make the system easier to use [13]. User comprehension of advice-giving systems is enhanced by transparency and scrutability, which allow the user to recognise when a system is inaccurate. These traits are often associated. Instilling trust in others is a well-known benefit of open communication.

Hybrid or neural-symbolic systems combine symbolic and sub-symbolic reasoning. The sub-symbolic system can develop predictive models using connectionist machine learning and analyse large quantities of data. In contrast, the extended system contains a rich representation of domain knowledge and can be utilised for higher-level, organised reasoning. The sub-symbolic components' assessments are described using these symbolic elements. The performance of an interpretable model may be evaluated by its accuracy and fidelity, whereas consistency and comprehensibility are desired attributes for the explainer. As a basis for common sense thinking, domain knowledge may be leveraged to aid with knowledge abstraction, refinement, and injection. When combined with a user's profile, a computer system can explain the sub-symbolic components and adjust the degree to which they are accurate and complex. It may also improve the sub-symbolic system's performance by fine-tuning and re-injecting the extracted information [14].

Machine Learning Model Post-hoc Explainability Techniques
The failure of ML models to meet any of the transparency criteria necessitates the development of a new technique that can be applied to the model to explain its results. Providing conveniently available information about how a model that has been previously developed produces predictions for each given input is the purpose of post-hoc explainability techniques. Several algorithmic techniques for post-hoc explainability are classified and discussed in this section, with a distinction made between (1) those developed for general applicability to ML models and (2) those developed for a single ML model and thus cannot be immediately extrapolated to any other learner. We will now go through the trends in post-hoc explainability for many ML models [16–18].

- Model-agnostic post-hoc explainability approaches may be used on any ML model without considering its internal processing or representations.
- Explainability after the customised fact or to explain specific machine learning models.

The deep learning techniques are convolutional neural networks, recurrent neural networks, and hybrid schemes encompassing both neural networks and transparent models that can be explained post-hoc and use layered structures of neural processing units. The most recent post-hoc ideas proposed by researchers for each model are thoroughly examined, and trend analysis is performed on these contributions [19].

Model-Independent Post-hoc Explainability Methodologies
Prototype-atheist post-hoc explainability approaches are intended to be put into any model to extract information from its prediction mechanism. Simplicity approaches are often used to develop proxies like their antecedents to create something tractable and straightforward. Sometimes the objective is to extract relevant information from the systems or visualise them to simplify understanding of their behaviour [20].

Model simplification, feature relevance estimations, and visualisation approaches may be used in prototype-atheist methodologies, as outlined in the taxonomy:

- Simplicity in explanation, the diverse methodology in model-independent post-hoc procedures. Because simplified models are often just representational of aspects of a model, this category also includes local explanations. Almost every strategy for model simplification that follows this route is based on rule extraction techniques. The Local Interpretable Model-Agnostic Explanations (LIME) method and its modifications are the most notable contributions to this approach. To explain an opaque model, LIME constructs locally linear models around its predictions. These contributions are classified as both simplified explanations and local explanations. G-REX is another way to rule extraction, in addition to LIME and similar flavours. Even though G-REX was not originally intended to be used for extracting rules from opaque models, the central premise of the algorithm has been extended to account for model explainability issues. A novel strategy for learning rules in CNF or DNF is proposed to bridge from a problematic model to a human-interpretable model, following rule extraction techniques, to be used in the future. Another study on the same issue describes model simplification as a model extraction approach that entails approximating a transparent model to a complicated model using a complex model as a starting point. An alternate technique to simplicity presented offers a way for distilling and auditing black-box models. The topics introduced are a strategy for model distillation and comparison to audit black-box risk score models and a statistical test to determine if the auditing data contains trained critical features. Because model simplification conforms to the most recent research on XAI, which includes methodologies such as LIME and G-REX [21, 22], it is easy to see why it is so popular. It seems from this that the post-hoc explainability technique on XAI will continue to be important shortly.
- To explain how an opaque model works, feature relevance explanation techniques attempt to rank or assess each feature's influence, relevance, or importance in the model's prediction output. This category covers a jumble of concepts, each of which employs a different computational approach to accomplish the same result as the previous one. In this regard, organisations such as SHAP, for example, have made substantial contributions (Shapley Additive Explanations). The authors devised a strategy for generating an additive feature significance score for each prediction with desirable qualities (local accuracy, missingness, and consistency) lacking in its predecessors. A further approach for addressing the contribution of each attribute to predictions is Coalitional Game Theory, which uses local gradients as a component. In a similar vein, local gradients are used to assess the modifications that must be made in each feature to change the model's output. When combined, several aspects that give insights into the data are grouped to allow the writers to evaluate the model's linkages and dependencies. There are many ways to measure how inputs influence system outputs presented in the study.

Shallow ML Models of Post-hoc Explainability

All kinds of supervised learning models fall under the umbrella of shallow machine learning. These models may be rigorously interpreted straightforwardly (Decision Trees and KNN). In contrast, more complicated learning techniques are used in other shallow ML models, necessitating different levels of justification. Because of ubiquity and substantial performance in prediction problems, tree ensembles and SVMs need post-hoc explainability procedures to justify their judgments [23].

Tree ensembles are undoubtedly one of the most accurate machine learning algorithms available today. They have developed a quick approach to increase the generalisation capabilities of single decision trees, which are prone to overfitting. This difficulty is avoided by using tree ensembles, which combine many trees to create an aggregated prediction or regression. While the combination of models is effective against overfitting, the interpretation of the complete ensemble is more challenging than the interpretation of each of its compounding tree learners, necessitating the employment of post-hoc explainability techniques. Many strategies are available for tree ensembles in the literature, including explanation by simplification and feature relevance approaches; numerous attempts have been made to reduce the complexity of tree ensembles while maintaining a part of the accuracy compensated for by the higher level of complexity. The author proposes that a series of random samples from the data (preferably matching the accurate data distribution) labelled by the ensemble model be used to train a single but less sophisticated model. Another method for simplification is described in [24], which involves the creation of a Simplified Tree Ensemble Learner (STEL). Similarly, using Expectation-Maximization and Kullback–Leibler divergence, [25] proposes using two models (uncomplicated and complicated), the former for interpretation and the latter for prediction.

In contrast to model-agnostic tactics, there are few options for obtaining board explainability in tree ensembles via model simplification, as there are for model-agnostic strategies. Consequently, either the methods presented are adequate, or model-agnostic approaches have already covered the whole range of simplification techniques. After simplifying the tree structure when dealing with tree ensembles, feature relevance techniques are used. A pioneering study conducted by Breiman et al. [26] investigated the importance of variables in Random Forests. He uses a random permutation technique to quantify the MDA (Mean Decrease Accuracy) or MIE (Mean Increase Error) of the forest when a given variable is randomly permuted in out-of-bag data. A Random Forest represents a complex system in a real-world situation. The importance of variables is used to reflect the underlying connections of the complex system in a real-world scenario. Following this author proposes a model for post-hoc explainability in which proposals are made that, if implemented, would convert an example from one class to another. This notion aims to break down the significance of the variables in a more transparent way. In this chapter, the authors explain how these tactics may be utilised to improve the effectiveness of risky online advertising while also improving their ranking in terms of payment rates.

Support Vector Machines

In the literature, the SVM is a well-known shallow machine learning model. As a result, the SVM method is more complex than tree ensembles, and their structure is far more complicated than Outlier detection. Other tasks may be accomplished by using an SVM to generate a hyper-plane in a high or infinite-dimensional space that can be utilised for classification. Hyperplanes furthest from the nearest training-data point of any class (so-called operating margin) tend to have better separation since larger margins reduce generalisation errors. Explanation through simplification, localisation, and visualisations are examples of post-hoc explainability in SVM applications. There are four types of simplifications used in explanation by simplification. The depth to which they delve into the algorithm's fundamental structure distinguishes them. Several writers suggest strategies for constructing rule-based models using a trained model's support vectors [29].

The strategy proposed in [30] uses a modified sequential covering algorithm to extract rules directly from the support vectors of a trained SVM. The authors also suggest eclectic rule extraction in [31], which takes a trained model's support vectors into account. Instead of traditional propositional rules, the work in [27, 28] creates fuzzy rules. Long antecedents diminish comprehensibility; hence, a more linguistically intelligible conclusion provides a more linguistically intelligible. The hyperplane, the support vectors, and the components responsible for building the rules are examples of the second class of simplifications discussed in the paper [32]. To generate hyperrectangles, he employs the intersections of the support vectors and the hyper-plane obtained by his technique. Another group of authors looked at including accurate training data as a component for developing the rules in a third way to model simplification, which was later abandoned. The authors could group prototype vectors for each class using a clustering method [33]. Their combination with the support vectors created ellipses and hyperrectangles inside the input space of this study. Additionally, the authors presented Hyper-rectangle Rule Extraction, a Support Vector Clustering (SVC) based technique for discovering prototype vectors for each class and then constructing small hyper-rectangles around them in a subsequent paper [35]. The rule extraction problem is a multi-constrained optimisation problem to build a collection of rules that do not overlap. Specifically, each rule denotes a nonempty hyper-cube with a shared edge with the hyper-plane and vice versa. For the same purpose, in a previous study, the authors developed a novel technique for extracting rules from gene expression data as a component of a multi-kernel SVM. This multi-kernel approach [36] incorporates features selection, prediction modelling, and rule extraction, among other things.

1.2   Intelligent Transportation Systems

Intelligent Transport System alludes to utilising data and correspondence innovations in transport. The advancement of ITS is yet developing. The degree to which these advancements are being used – and the complexity in their arrangement – differs starting with one country and then onto the next. Transport experts throughout the planet need to comprehend the central applications and capacities so they can evaluate potential benefits, related expenses, and how ITS might best be conveyed. ITS Security is fundamental because of the progression and development of new advancements in transportation. The weakness in the advances makes it more straightforward for the aggressor to upset the ITS. Various transportation methods also traffic the executives to utilise ITS's inventive administrations for more astute transportation organisations. ITS are urgent to make shrewd urban communities, and assaults to these frameworks can cause genuine circumstances for the transportation inside a city. ITS are a significant piece of people's vehicles for security and extravagance. ITS consolidates a broad scope of utilisations, and it keeps on expanding the significance of individual and public transportation frameworks. These applications cycle and share the data to upgrade traffic stream, executives' traffic, and ecological effects from transportation frameworks [37].

Preparation is an essential part of transportation planning since it identifies and addresses community needs while considering transportation's social, environmental, and economic aspects. Part of the Network Design Problem (NDP) involves designing an optimum road strategy for transportation planning. In certain cases, it may be both a Continuous and a Discrete issue, depending on how the capacity of the current infrastructure evolves. Neural Networks for street development, design, and modelling were previous studies' topics in the 1990s. According to reference [39], a parallel neural network system simulates the spatial link between transportation and land-use planning. To determine the most efficient route in urban planning, researchers turned to raster algorithms, which don't need existing linkages and nodes. Most studies now focus on most studies because of the enormous quantity of data and powerful algorithms generated. Using a bi-non-linear type task in conjunction with two stages tackled the ongoing NDP issue [38]. They constructed a network and compared the performance of GA and SA algorithms. When demand is minimal, SA is more efficient than GA at finding the ideal value. When GA carries out more calculations, it may find a better answer. "According to GA produces greater outcomes than SA in less time"; however, this study shows that this is not the case. It was just a single-level linear model for the Continuous NDP issue that this research examined. Abduljabba et al. [39] suggested ANN and genetic algorithms mimic the city of Ankara, Turkey's safety management plan. The ANN model's findings were more accurate than GA's, and there was less room for mistakes. Based on the simulated industrial land-use patterns over several years in China, the findings demonstrated an improvement in planning for urban growth. There is a strong transportation system management and safety plan and a correct allocation of roadways, trains, and airways [39]. Predicting accidents and injuries on the Istanbul-Ankara highway was done using an artificial neural network (ANN). Compared to GA and AIS, SA was less effective in this situation. Congestion and long wait times may be avoided by carefully designing vehicle routes. The ant colony method is a potential solution to the issue of vehicle routing [34, 38].

In contrast, they are concentrating on applying the BCO Algorithm to solve a routing and wavelength issues. This challenge is all about selecting a network route and allocating wavelengths to linked nodes to maximise the number of connections between the nodes. The use of microscopic traffic data to simulate and detect security vulnerabilities and traffic control systems and road management plans has also been studied recently [40]. It proposes the optimal routes for public transportation customers, and the real-time path creation system should be taught and updated based on travellers' preferences. A utility-based method examines various qualities of pathways and parameters for each public transportation user. Intelligent Transportation Systems (ITS) is another area where AI applications have experienced tremendous progress (ITS). These systems use several technologies and communication networks to reduce traffic and enhance the driving experience. They collect critical information that may be used in machine learning algorithms. Deep reinforcement learning, for example, has been utilised to optimise traffic management rules in large ITS systems in real-time [41]. It has also been suggested that ITS devices be equipped with signal processing and rapid computing analytics via a deep learning system [42]. A more connected transportation system will need deep learning methods in the future because of the increasing complexity of the data that ITS generates. In another instance, traffic light systems at intersections were automatically controlled using evolutionary algorithms and fuzzy approaches. According to research, an RFID-based traffic management system called 'NeverStop,' was shown to cut vehicle wait times. Two NNs systems were designed to control better the road based on microscopic simulation data. Controlling traffic signals and predicting traffic congestion are two separate functions of the first system. The multi-layer NNs system proposed in three intersecting networks showed the possibility of utilising NNs to regulate traffic [39]. Control of signal traffic may also benefit from the usage of ANN-based algorithms. Two NN systems based on tiny, simulated data were developed to control the road better. The first is a traffic signal control system, while the second is a traffic congestion prediction system. A multi-layer NNs system tested in three intersecting networks revealed the viability of utilising NNs to regulate traffic [39].

Reward learning neural networks (NNs) are used to update the system's parameters and cycle lengths as the flow of data changes. New approaches and applications are constantly being developed to make the most of AI's potential for enhancing road planning, decision-making, and management.

How Will It Change Our Lives in the Future?
With the developing accessibility of sensors and observing gadgets – all pieces of the Internet of Things pattern – more enormous datasets will stream into the control place, and the apparatuses used to screen and deal with this information will thrive. Transportation frameworks can profit from these overall patterns, presenting more practical improvements. Similarly, the cell phone has altered the dissemination of traveller data.

1.3   AI and Infrastructure to Support ITS

Information and communication technologies play a significant job in transportation and traffic the board frameworks. Consolidation of insight into transportation and traffic the board frameworks is expected to work on the well-being, productivity, and maintainability of transportation organisations, decrease gridlock and improve drivers' encounters. The mix of insight, data, and correspondence innovations bring about Intelligent Transportation Systems (ITS); as these extra advances get added to the board frameworks' transportation and traffic, the surface region for the assault increments. Consequently, the need to secure the ITS increments [42].

As the volume and thickness of vehicles increments, innovative headways have grown better approaches to overseeing traffic. The Intelligent Transportation Systems applies these innovative advances to street transport. Data is gathered from sensors and hardware in vehicles and foundations. The data can be utilised to develop the current transportation frameworks further. Street and traffic security, traffic proficiency, and worth-added applications point to work on the transportation framework.

Transport Industry has been a significant supporter of developing individuals and merchandise across different geological districts. It assumes a huge part of the board framework's store network, where inventory moves from one spot to the next. The business accepts a critical component in the development of merchandise to the perfect convergence of everything working out in a coordination chain. To receive the total reward from business speculation, advances like Machine Learning, Artificial insight, and the web of Things have been utilised by legislatures and associations.

Most of the enormous urban communities across the globe face issues identified with transport, traffic, and coordination. Because of the quickly developing human populace and the expansion in the number of vehicles out and about. To proficiently make and deal with a feasible vehicle framework, innovation could be of tremendous help. With metropolitan regions battling gridlock, AI arrangements have arisen to get consistent data from vehicles for traffic the executives and use versatility on request in trip arranging through a solitary user interface. A safe mix of AI-based direction, traffic the board, steering, transportation network administrations, and other versatility enhancement instruments are different conceivable outcomes of proficient traffic the executives.

On the other hand, the integrated technique does not mimic individual artificial intelligence (AI) applications on a case-by-case basis. Instead, a general frame model with many substructures is built, i.e., an AI toolbox with each AI tool in a visible form. As a result, the AI framework idea offers a general superstructure for combining different AI systems. This strategy has the advantage of emphasising causal and structural connections and introducing new ideas and research discoveries to the whole structure. The integrated AI Framework can also explain the important things to be studied – fundamental factors, constructs, or variables – in either graphic or narrative form, as well as the presumed interrelationships between, easing management and guidance in a complex and new research field like AI-assisted customer services. Because of this abstract superstructure, the technical correctness and qualities of the used model are required. The model frame's abstraction prevents sub-sections from being expanded and presented precisely, allowing part of the original data and results to be excluded during the transition between the individual and ideologically homogeneous perspectives [43].

As a result, the AI Framework Model must balance clarity, creativity, and abstraction. However, the focus is on the business model element and the supervisory body, setting standards and restrictions on using AI in public service and support services. The framework model also describes major technical processes in Fig. 1.3 and ways of operating to deal with the lack of technical accuracy of the integrated perspective.

1.4 Overview of ITS Applications

ITS is recognised as a high potential space to handle the many difficulties confronting the Transport area. Indeed, other than a framework, the ITS is considered the single generally significant "factor" that can impact participation among the different methods of moving and make a consistent transportation framework across Europe. Today, there is also a substantial and various partner local area which is either giving or using ITS applications and administrations. The area accommodates a significant commitment to monetary and social turn of events.

The organisation of ITS frameworks and administrations has been generally "unimodal" in degree and degree, leaving the more extensive "cross modular" utilisation of current frameworks behind and further advancement to be looked for later. Likewise, the turn of events and use of unimodal ITS applications are seen by numerous individuals as still fragmented and not broad enough to cover the entirety of typical applications and market take-up extensively. Accomplishing the "minimum amount" for self-adequacy and manageability of incorporated ITS applications is still an objective [44].

Endeavours to improve ITS multiplication at the National and EU level have been strengthened in the last decade through both administrative and specialised advancement measures. The soonest "institutional" endeavours to advance ITS in Europe came in the mid-'90s. From that point forward, continuously and consistently, ITS frameworks and applications are constantly being created and carried out in Europe and all through the developed world. In 2008 the Commission gave a Communication on an "Activity Plan for the Deployment of Intelligent Transport Systems in Europe." It has required an appraisal of the approach needs, the decision of conventional ITS parts to be shared or re-utilised, and settlement on a reasonable schedule for execution under the accompanying "Activity regions":

• Traffic and transport management ITS services on European corridors and in conurbations.

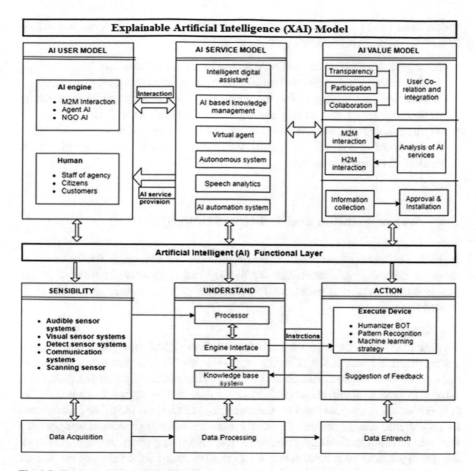

Fig. 1.3 Framework of Explainable AI

- Road security and stability.
- Vehicle integration into public transportation systems.
- Data security measures, as well as liability concerns.
- Coordination and integration of European ITS systems.
- To make the best possible use of traffic, road, and travel data.

In the 2010 directive, the accompanying need regions are outlined for the turn of events and utilisation of determinations and norms to give interoperability, similarity, and congruity for the organisation and functional utilisation of ITS:

- Ideal utilisation of street, traffic, and travel information,
- Progression of traffic and cargo the executives ITS administrations,
- Street well-being and security applications, connecting the vehicle with the vehicle foundation.

Inside these need regions, various need activities for the turn of events and utilisation of ITS details and principles are set as follows:

- Arrangement of EU-wide multimodal travel data administrations.
- Arrangement of EU-wide continuous traffic data administrations.
- Information and strategies for the arrangement, where conceivable, of street well-being related least widespread traffic data for nothing to clients.
- Orchestrated arrangement for an interoperable EU-wide e-Call.
- Arrangement of data administrations for free from any harm leaving places for trucks and business vehicles.

1.5 XAI in Aviation and Transportation System

Field data collecting has become more dependable and accessible due to advancements in computer science. Given the density and unstructured nature of public transportation data, it is necessary to model the data effectively to get the most benefit from it.

Aviation
The use of artificial intelligence has been shown to improve the efficiency of flight management. AI may assist in technology (ML and DL), software/hardware, and application development. Parachute Low Altitude Delivery System (PLADS) was created in 1989 to obtain evidence from very impenetrable aircraft records and alter it to enable the SA method and vector machine, as described [45]. They discovered that SVM produces satisfactory results for this form of categorisation. Unsupervised machine learning methods may be used to improve safety when an airplane is landing, and a probabilistic neural network was used to evaluate the safety of the aircraft by examining the engine on-board, which was believed to be rather sophisticated at the time of PNN. In terms of training the network, it is identical to MLP, except that it employs a radial basis function rather than a linear basis function. As a result, more neurons are needed in the hidden layer of RBN than in MLP. A further development was developing an automated supervised Random Forest approach for identifying aeroplane turbulence with higher accuracy than previously available methods. In addition to assisting the pilot in avoiding straying from the predetermined course, this technology may also aid the pilot in minimising fuel consumption and enhancing air control management [45].

Shared Transport
The sharing economy in mobility has proven attractive to reducing traffic and pollution in crowded metropolitan areas because of its capacity to propose solutions to remove single or fewer occupant cars from the road. It is a win-win strategy since this approach's short-term socio-economic and environmental advantages may be reaped while long-term sustainable solutions can be implemented. "ICT-enabled platforms for exchanging goods and services relying on non-market logics such as

sharing, lending, giving and swapping, and market rationale; renting and selling" define the sharing economy. Uber and Airbnb are leading the charge in the sharing economy industry in the transportation and lodging industries. There is a potential interest in establishing new business models for new-shared mobility services in the transportation industry to create a sustainable transportation system and fill the gap between demand and supply [46]. In particular, on-demand bike, car, and ridesharing have captured the interest of the bulk of consumers. These shared mobility choices also address the "first-mile, last-mile" problem.

Sharing mobility options like bike and automobile sharing have been around for a long time. Disruptive shared mobility ideas may now be implemented primarily because of information and communication technology (ICT). Many apps allow for an efficient transportation system by coordinating on-demand automobiles and linking them with people who want to share trips. AI technology is helping to enhance the customer experience and simplify operations in the shared transportation sector. Combining AI with shared mobility has made it feasible to provide a tailored consumer experience to users [47]. Uber, for example, provides a tailored passenger experience by recommending locations based on a user's previous rides. Route-based pricing, which employs AI to estimate how much passengers are ready to pay depending on destination, time of day, and location, has also been adopted by Uber to serve its customers better. Uber's operators are also using AI to detect and prevent fraudulent behaviours by drivers. Driven by AI and sensor technology, driverless vehicles are the future of shared transportation. Autonomous fleets are more lucrative, and by 2030, they might replace 6.2 million drivers throughout the world. Sharing-transit companies like Uber, which rely on AI to design routes and precisely estimate trip times, have also benefited from these improvements in AI in transportation. As a result, public transit will be more dependable and efficient. In addition, big data and ANN have made it feasible to estimate passenger demand for public transit based on prior trip data and environmental information. An algorithm created by Ma et al. [48] targets prospective passengers during reservations when buses are idle and suggests where they should go. An excellent pick-up point selection technique has been discovered using machine learning information via Q-learning [47].

Public Transport

Several studies have been carried out to improve the safety and reliability of bus rides and destinations using a hybrid Ant Colony Algorithm (HACA). Artificial neural networks (ANNs) to anticipate bus arrival timings may also cut passenger wait times. Automatic buses are another use of this technology. For the first time, the iBus architecture was introduced. It goes through the same three stages as a human driver: perception, decision-making, and action, but using software and technology instead. The iBus was put through its paces in the People's Republic of China. Alpha Buses is another kind of autonomous bus in China. Singapore has an agreement in place for an automated shuttle service to be tested on the island of Sentosa in 2022; commuters in the districts of Punggol, Tengah, and the Jurong Innovation District (JID) will be able to take advantage of the Smart bus service in Singapore as well [49]. It is even more impressive to introduce a self-driving Olli bus that carries passengers to their destinations and communicates with them about the route, surroundings, and Olli itself. Demand responsive public transportation with AI developments such as flexible on-demand bus services that run on flexible schedules and itineraries are another standout. Taxi-like convenience and bus-like efficiency are the goals of these services, striving to combine the best of both worlds. Efforts are being made across the globe to increase traditional bus services' efficiency and give passengers more convenient options. Optibus has released a new AI-driven On-Time optimisation solution to help transport operators eliminate delays and offer on-time service. In China, a demand-responsive customised bus (CB) has been implemented for public transportation. 2014 saw the inauguration of the Boston-based BRIDJ smart, flexible bus service. Our system uses real-time traffic big data and passenger inputs to determine how and where people desire to go. The programme can identify the quickest route and only stops at sites requested by passengers to maximise service. To satisfy your customers, you must give fast and effective service. In addition, it is critical to keep operating costs down by ensuring that the fleet size is appropriate and that the occupancy level is at its optimal level. The passenger demand forecast is critical to determining the bus headway, a fleet size optimally used by the fleet. A traditional bus system's passenger demand forecast models may be a starting point. Another area where artificial intelligence has shown its effectiveness is monitoring vehicles on transportation networks. An Automatic Vehicle Location (AVL) system is installed to improve operating efficiency, manage operational control, and improve the overall quality of public transportation services. This system uses GPS signals to extract information to track transportation units in real-time, diagnose defects, and notify vehicles of any changes. Additionally, travellers may get information on their mobile devices via application programming interfaces (APPs). The installation of an AVL-based GPS in Cagliari, Italy, was a success. Every 30 seconds, AVL receives real-time information on the location of buses from the bus company. These statistics are used for three different purposes: to enhance bus service reliability, prioritise bus movement at traffic signals, and educate passengers about the schedule of the next bus near bus stop locations. Another example of an AVL-based GPS is the iBus system in London. In addition, artificial intelligence-based data management solutions have been created to ensure the successful implementation of AVL systems [50]. Data from the automated vehicle location system (AVL) was combined with machine learning clustering techniques to enhance the AVL system's performance for scheduling bus routes in Portugal. The integration was utilised to analyse whether a schedule alteration would meet the needs of the network in question. We can improve public transit schedules in Sweden by analysing AVL data using unsupervised clustering machine learning approaches. BusGrid leverages data from the AVL to estimate future passenger demand at bus stops and routes using supervised machine learning algorithms. It helps to enhance the design of new routes and the scheduling of existing bus routes. The data revealed that the algorithm could effectively anticipate the demand for bus transportation. There are various issues with AVL-based public transit planning and research gaps. There is a lack of dependability evaluation for current Schedule Planning (SP) indicators by monitoring the frequency on each route, which would allow for more accurate scheduling. It is also vital to determine whether SP can accommodate historical system behaviour while still meeting current demand patterns [47, 48]. A second hole is an assessment of utilising ANN to improve regression issues based on travel time prediction and improve the arrival schedule for buses using regression models and speed based Kalman filtering. The last difficulty is implementing an automated control approach employing AVL and APC data for rapid reaction to events and successful public transportation planning. The smart card data may be used better to meet the needs and preferences of public transportation users. Passengers are the focus of most smart card data, and hence, numerous studies have been undertaken on guessing passengers' final destinations from this data. It uses deep learning as one of these methods. Validation, size of the sample, information (the calibre of methods), sensitivity analysis, and other variables that might influence the suggested models are all considered while assessing various approaches. According to the authors, there is a lack of validation and sensitivity analysis in most studies. As a bonus, they recommended including land-use variables and information to improve the present model on nearby public transportation systems. It is possible to predict a passenger's final destination using GPS and video data from many sources [51].

Urban Intelligent Mobility
Optimal utilisation of infrastructure and better decision-making based on real-time data are the ambitions for the future of intelligent urban transportation. Furthermore, creating an intellectual connection to establish a sustainable, seamless, and environmentally-friendly network benefits people and the environment. Autonomous vehicles (AV), or cars that can operate independently of a human driver, were recently becoming a reality. Deep learning methods are used to develop AI software for autonomous vehicles (AV). By teaching the car how to drive while preserving safe headways, lane discipline, and control of the vehicle, this strategy works [52]. The influence of autonomous vehicles (AVs) on traffic safety and congestion and the possibility of shifting travel habits have been extensively studied and projected. People's travel habits are predicted to alter due to these AVs, leading to new social structures and urban designs. New business models will aid in vehicle sharing and ridesharing with innovative solutions to current hurdles, such as restricted accessibility and dependability. Liu et al. [55] explained the origins of self-driving cars. A "urbmobile" first electric car is developed by The Cornell Aeronautical Laboratory (CAI) in 1968; however, the technology didn't promote its use because of its high cost. Due to the increasing availability of sensors and cameras over the last several years, it is now more essential to build more dependable and safer roadways by completely automating the cars. It is said that recognising traffic signals, other cars, road signs, and weather conditions is difficult for a vehicle operating in metropolitan areas because of the weather and lightning. The security and integrity of the system and software difficulties are critical to this new technology. Hardware and software architecture make up the bulk of these self-driving cars. Computer systems and actuators make up the hardware architecture, while a localisation technique and

an ability to identify moving objects make up the software component. Sensor-based technology and inter-vehicle communication are at the heart of AVs. In the future, this AV will be able to operate entirely autonomously, avoiding obstructions and people on the road while still being able to drive itself. This capacity is produced using artificial intelligence, pattern recognition algorithms, sensors, and 3D cameras.

Google unveiled an autonomous Toyota Prius in the United States in 2010. Over 30,000 lives are expected to be saved, and the annual cost of traffic accidents in the United States is anticipated to be 270 billion dollars. The vehicle's ability to self-park will further reduce the need for parking lots. Lexus RX450h and "Firefly" were both unveiled by Google in 2012 and 2015, respectively, as fully autonomous vehicles with no steering. Google has formed a separate business called "Waymo" to continue developing its driverless car project. According to the client, approval faces many challenges, the most significant societal acceptability. AV has performed worldwide research on a distinct component of making the intelligent cognitive vehicle more attractive to passengers [53].

The following were the main points of discussion:

- Automobiles can remedy their own mistakes. Vehicles capable of self-socialising can communicate with other vehicles and people in natural language.
- Self-learning: The vehicle learns from its actions, as well as those of the driver, passengers, and the environment at large.
- Vehicles that can drive themselves autonomously in a controlled setting are self-driving.
- The digital information in each mobility may determine each vehicle's desired and tailored experience. Like other intelligent transportation gadgets, self-integration is a crucial feature of this technology.

ITS for Traffic and Travel Information (TTI)
Traffic and Travel Information (TTI) advancements, frameworks, and enacting arrangements focusing on giving constant data to the explorers were the main parts of ITS to be created and advanced industrially. TTI includes the EU's ITS Action Plan, the ITS Directive, and the CEN-ETSI reaction to Mandate M/453 on Co-employable Systems Normalization. A definitive degree and objective of TTI frameworks are to: give ceaseless and dependable traffic and travel information and data of significance to all modes and organisations through general admittance to such data and information trade (across locales and boundaries), empowered by practical plans of action [54].

TTI is accordingly about the assortment, handling, communication, and ideal utilisation of Traffic and Travel information for provincial, National, and EU-wide continuous travel data, and its arrangement (a base of which ought to be free) to clients of such data accordingly empowering the improvement of different business esteem added administrations. In evaluating the analytical and formative capability of TTI advances and administrations, there are four principal issues to be thought of, for example:

- Legitimate structure for the arrangement of TTI,
- Specialised principles utilised and the interoperability of these norms,
- Plans of action used for the arrangement of this information, and
- The degree to which the actual voyagers (for example, the "request") respond to the TTI arrangement.

ITS for Traffic and Public Transport Management

Street traffic light frameworks are considered as the foundation of Intelligent Transport Systems, in the feeling of enormous scope, region-wide framework executions, managing a lot of traffic (also other) information put away and handled progressively from different sources/locators, with the utilisation of progressed traffic models, expectation calculations and the executives' systems ready to react in continuous to the prevailing traffic conditions. Some critical advances for the up-and-coming age of traffic the board frameworks inside enormous scope execution of ITS are traffic prescient control techniques and intelligent organisation control systems dependent on specific winning rules [56].

- Intelligent control of traffic signals.
- Identification and management of incidents.
- Priority should be given to emergency and public vehicles.
- Intelligent lane control.
- Enforcement of speed restrictions.
- Diversions across longer distances, re-routing.
- Compilation of Information.

An examination of ITS for-street traffic board upheld in the EU since the 1980s, yet even more thoroughly from the mid-1990s with the Euro-Regional undertakings, to further develop traffic the board and client administrations, cantering in crossline hallways. Simultaneously, projects under the CIVITAS drive carried out and shared metropolitan vehicle and traffic arrangements, including light traffic administration, the need for public vehicles, and the extensive portability of the board approach [55–57].

1.6   Intelligent Transportation Systems Past, Present, and Future

The edge of advanced mechanics has never been characterised however is more considerable than data and interchanges advancements. An advanced mechanics framework is an innovative substance that has discernment, choice, and activity and at last correspondence functionalities, and which interfaces with an actual climate to supplant absolutely or somewhat an individual. The jug neck of advanced mechanics is by an extensive insight and circumstance examination [50].

According to our assumptions about transportation frameworks, a vehicle may be seen as a flexible robot with lively independence but has not yet achieved its decisional autonomy. There are already driverless trains and tramways in operation in several countries. The high-level driver assistance frameworks that are already in place will combine in the next 10 years to create a driverless car under specified scenarios, as shown in the example above. Since there are around one billion automobiles on the road globally, digital vehicles may be the next wave of portable robots [58].

In every critical country around the globe, the automobile industry is an A-list sector responsible for driving economic progress. The vehicle sector is still in growth mode, as seen by its impressive 30%+ growth rate during the last decade. The global automobile industry produces over 70 million automobiles, vans, trucks, and transporters every year. These vehicles are critical to the operation of the worldwide economy and the prosperity of the people. Automobiles, trucks, transportation, and mentors are estimated to generate more than $50 million in revenue throughout the globe and are a significant street organisation to convey protected and smooth traffic [59].

The main three issues are as follows:

* Energy utilisation
* Well-being
* Versatility and clog

Every year, around 1.2 million people are killed or injured on the streets of the globe. However, if we look at the countries where those deaths happened, we can see that only around 15% of tragic events occur in wealthy countries, whereas 85% occur in impoverished ones. For example, in Europe or North America, there are "about" 50,000 deaths per year, but in India, there are 220,000 fatalities per year. It creates a Catch 22 situation in which deadly accidents occur in areas with few cars, but there are fewer in areas with numerous vehicles. The explanation may be found in the impotent street network, the maturing armada, and the lack of driver training [58, 59].

Despite this oddity, it is accurate to proceed with research on further developing security. Over 90% of mishaps is due to human disappointment. So, the halfway or all-out robotisation of driving that at last limits or even kills people subsequently dispenses with chances. Driver well-being may be enhanced significantly with the help of Advanced Driver Assistance Systems (ADAS). They intend to investigate some of the breakthroughs in data and communication that have emerged in recent years.

Advanced Driver Assistance Systems (ADAS)

ADAS's objective is to improve security and the versatility of street traffic. A massive assortment of ADAS has been developed in the research centres, and a few of them are monetarily accessible.

A non-thorough rundown is the accompanying:

* Adaptative Cruise Control (ACC)
* Path Departure Warning System
* Progressed Navigation System

- Versatile light control
- Vulnerable side Detection
- Driver Drowsiness Detection
- V2V for vehicle-to-vehicle correspondence
- V2I for the vehicle to Infrastructure correspondence
- Crisis Call (eCall)

ACC is an advanced driver assistance system that uses either a radar or laser setup to allow the car to slow down while approaching another vehicle and accelerate back up to the current speed when traffic permits. The cost of laser-based frameworks is much cheaper than that of radar-based frameworks. Despite this, laser-based adaptive cruise control frameworks do not reliably detect and follow cars in less-than-ideal weather circumstances, nor do they track highly dirty (non-intelligent) vehicles. Some frameworks also feature forward impact cautioning or crash moderation aversion systems, which warn the driver and apply brake pressure if they believe there is a significant risk of imminent rear-end collision.

Lane Departure Warning System (LDWS)

LDWS is an advanced driver assistance system (ADAS) designed to warn a driver when a vehicle begins to travel off its planned course unless the vehicle's blinker is turned on in the direction of the intended path. LDWS is most often used on interstates and arterial roads like blood vessel streets. Tiers developed the first LDWS framework in Europe for Mercedes Actros business trucks, the first of its kind globally. It is now available on trucks in North America and Japan. In these frameworks, a detectable thunder strip sound is produced on the vehicle drifting out of the path, alerting the driver to the possibility of an inadvertent path take-off. When the motorist uses the blinker, the system recognises that they intend to exit the road, but no notifications. Citroen first introduced LDW on their C4 and C5 cars in 2005, and they are now offering it on their C6 models throughout Europe. These sensors, located beneath the front guard, are used to filter route marks out and about on the ground surface using this framework. A vibration instrument in the driver's seat alerts them of potential deviations [48–51].

On-Board Signaling

The route is created by combining a precise limitation framework that uses GPS and odometers with a guide that coordinates activities and places the car on a computerised map. It is true that the advanced map is, in fact, a chart and that the route makes use of a directed calculation between the starting stage and the appearance point, like that which is used in electronic circuits. An example of a street data set is a vector guide of a particular area of interest. Road names or numbers and home numbers are encoded as geographic coordinates to allow the client to locate an ideal target using the road address. Focal points will also be labelled and stored with their geographic coordinates. Speed limit signs, street surveillance cameras, gasoline stations, inns, eating stops, and hospitals are focal areas in a neighbourhood. We refer to this as the original or static route since we are just considering the street network's computation and not the traffic's thickness [60–64].

1.7   Conclusion

The chapter focuses on XAI, a fundamental necessity for adopting machine learning algorithms in real-world applications. As a result of our work, we've gained new insight into the topic of model explainability and the many factors that drive researchers to seek out more easily understandable machine learning algorithms. The recent literature on explainability provides a solid foundation for two approaches: (1) ML models, making them interpretable to some extent by themselves, and (2) Explainable AI techniques devised to make ML models interpretable. The literature analysis has produced a global taxonomy of numerous proposals made by the public. ITS intends to improve safety, comfort, transportation efficiency, and mitigate transportation's environmental consequences; however, all these goals may be diluted in the presence of security footholds. ITS availability, authentication, integrity, privacy, and security objectives may be exploited with any weaknesses. Furthermore, numerous problems must be answered before a large-scale automotive ad-hoc network can be implemented.

References

1. S.J. Russell, P. Norvig, Artificial intelligence: a modern approach, Malaysia; Pearson Education Limited 2016.
2. A. Preece, D. Harborne, D. Braines, R. Tomsett, S. Chakraborty, Stakeholders in Explainable AI, 2018.
3. Gaur, L., Bhandari, M., Razdan, T., Mallik, S., & Zhao, Z. (2022). Explanation-driven deep learning model for prediction of brain tumour status using MRI image data. Frontiers in Genetics, 13 doi:https://doi.org/10.3389/fgene.2022.822666
4. L. Gaur, U. Bhatia, N. Z. Jhanjhi, G. Muhammad, and M. Masud, "Medical image-based detection of COVID-19 using Deep Convolution Neural Networks," Multimedia Systems, 2021, doi: https://doi.org/10.1007/s00530-021-00794-6.
5. J. Zhu, A. Liapis, S. Risi, R. Bidarra, G.M. Youngblood, Explainable AI for designers: A human-centered perspective on mixed-initiative co-creation, 2018 IEEE Conference on Computational Intelligence and Games (CIG) (2018) 1–8.
6. F.K. Došilović, M. Brčić, N. Hlupić, Explainable artificial intelligence: A survey, in 41st International Convention on Information and Communication Technology, Electronics, and Microelectronics (MIPRO), 2018, pp. 210–215.
7. P. Hall, On the Art and Science of Machine Learning Explanations, 2018.
8. Arrieta, Alejandro Barredo, Natalia Díaz-Rodríguez, Javier Del Ser, Adrien Bennetot, Siham Tabik, Alberto Barbado, Salvador García et al. "Explainable Artificial Intelligence (XAI): Concepts, taxonomies, opportunities and challenges toward responsible AI." Information Fusion 58 (2020): 82–115.
9. T. Miller, Explanation in artificial intelligence: Insights from the social sciences, Artif. Intell. 267 (2019) 1–38.
10. L.H. Gilpin, D. Bau, B.Z. Yuan, A. Bajwa, M. Specter, L. Kagal, Explaining Explanations: An Overview of Interpretability of Machine Learning, 2018.
11. A. Adadi, M. Berrada, Peeking inside the black-box: A survey on explainable artificial intelligence (XAI), IEEE Access 6 (2018) 52138–52160.

12. O. Biran, C. Cotton, Explanation and justification in machine learning: A survey, in IJCAI-17 workshop on explainable AI (XAI), 8, 2017, p. 1.
13. S.T. Shane, T. Mueller, R.R. Hoffman, W. Clancey, G. Klein, Explanation in Human-AI Systems: A Literature Meta-Review Synopsis of Key Ideas and Publications and Bibliography for Explainable AI, Technical Report, Defense Advanced Research Projects Agency (DARPA) XAI Program, 2019.
14. Gunning, David, and David Aha. "DARPA's explainable artificial intelligence (XAI) program." AI Magazine 40, no. 2 (2019): 44–58.
15. R. Guidotti, A. Monreale, S. Ruggieri, F. Turini, F. Giannotti, D. Pedreschi, A survey of methods for explaining black box models, ACM Computing Surveys 51 (5) (2018) 93:1–93:42.
16. G. Montavon, W. Samek, K.-R. Müller, Methods for interpreting and understanding deep neural networks, Digital Signal Processing 73 (2018) 1–15, doi: https://doi.org/10.1016/j.dsp.2017.10.011.
17. A. Fernandez, F. Herrera, O. Cordon, M. Jose del Jesus, F. Marcelloni, Evolutionary fuzzy systems for explainable artificial intelligence: Why, when, what for, and where to? IEEE Computational Intelligence Magazine 14 (1) (2019) 69–81.
18. R.S. Michalski, A theory and methodology of inductive learning, in: Machine learning, Springer, 1983, pp. 83–134.
19. D. Doran, S. Schulz, T.R. Besold, What does explainable AI really mean? a new conceptualisation of perspectives, 2017.
20. A. Vellido, J.D. Martín-Guerrero, P.J. Lisboa, Making machine learning models interpretable., in: European Symposium on Artificial Neural Networks, Computational Intelligence and Machine Learning (ESANN), 12, Citeseer, 2012, pp. 163–172.
21. A. Holzinger, C. Biemann, C.S. Pattichis, D.B. Kell, What do we need to build explainable AI systems for the medical domain?, 2017.
22. M.T. Ribeiro, S. Singh, C. Guestrin, Why should I trust you? Explaining the predictions of any classifier, in: ACM SIGKDD International Conference on Knowledge Discovery and Data Mining, ACM, 2016, pp. 1135–1144.
23. H.C. Lane, M.G. Core, M. Van Lent, S. Solomon, D. Gomboc, Explainable artificial intelligence for training and tutoring, Technical Report, University of Southern California, 2005.
24. W.J. Murdoch, C. Singh, K. Kumbier, R. Abbasi-Asl, B. Yu, Interpretable machine learning: definitions, methods, and applications, 2019.
25. J. Haspiel, N. Du, J. Meyerson, L.P. Robert Jr, D. Tilbury, X.J. Yang, A.K. Pradhan, Explanations and expectations: Trust building in automated vehicles, in: Companion of the ACM/IEEE International Conference on Human-Robot Interaction, ACM, 2018, pp. 119–120.
26. Gaur L, Singh G, Solanki A, Jhanjhi NZ, Bhatia U, Sharma S, et al. Disposition of youth in predicting sustainable development goals using the neuro-fuzzy and random forest algorithms. Hum Cent Comput Inf Sci. (2021) 11:24. doi: https://doi.org/10.22967/HCIS.2021.11.024
27. A. Chander, R. Srinivasan, S. Chelian, J. Wang, K. Uchino, Working with beliefs: AI transparency in the enterprise., in: Workshops of the ACM Conference on Intelligent User Interfaces, 2018.
28. AB Tickle, R. Andrews, M. Golea, J. Diederich, The truth will come to light: Directions and challenges in extracting the knowledge embedded within trained artificial neural networks, IEEE Transactions on Neural Networks 9 (6) (1998) 1057–1068.
29. C. Louizos, U. Shalit, J.M. Mooij, D. Sontag, R. Zemel, M. Welling, Causal effect inference with deep latent-variable models, in: Advances in Neural Information Processing Systems, 2017, pp. 6446–6456.
30. Gaur Loveleen, Bhandari Mohan, Bhadwal Singh Shikhar, Jhanjhi Nz, Mohammad Shorfuzzaman, and Mehedi Masud. 2022. Explanation-driven HCI Model to Examine the Mini-Mental State for Alzheimer's Disease. ACM Trans. Multimedia Comput. Commun. Appl. (March 2022). doi:https://doi.org/10.1145/3527174

31. Gao, Jun, Ninghao Liu, Mark Lawley, and Xia Hu. "An interpretable classification framework for information extraction from online healthcare forums." Journal of healthcare engineering 2017 (2017).
32. Harradon, Michael, Jeff Druce, and Brian Ruttenberg. "Causal learning and explanation of deep neural networks via autoencoded activations." arXiv preprint arXiv:1802.00541 (2018).
33. Hefny, Ahmed, Zita Marinho, Wen Sun, Siddhartha Srinivasa, and Geoffrey Gordon. "Recurrent predictive state policy networks." In International Conference on Machine Learning, pp. 1949–1958. PMLR, 2018.
34. Mathur, S., & Gaur, L. (2021). Predictability, power and procedures of citation analysis doi:https://doi.org/10.1007/978-981-15-9689-6_6.
35. M. Chaudhary, L. Gaur, N. Z. Jhanjhi, M. Masud, and S. Aljahdali, "Envisaging Employee Churn Using MCDM and Machine Learning", Intelligent Automation & Soft Computing DOI:https://doi.org/10.32604/iasc.2022.023417
36. Hoffman, Robert R., and Gary Klein. "Explaining explanation, part 1: theoretical foundations." IEEE Intelligent Systems 32, no. 3 (2017): 68–73.
37. Abduljabbar, Rusul, Hussein Dia, Sohani Liyanage, and Saeed Asadi Bagloee. "Applications of artificial intelligence in transport: An overview." Sustainability 11, no. 1 (2019): 189.
38. D. K. Sharma, L. Gaur, and D. Okunbor. Image compression and feature extraction using kohonen's self-organising map neural network. Journal of Strategic E-Commerce, 5:25–38, 2007.
39. Abduljabbar, Rusul, Hussein Dia, Sohani Liyanage, and Saeed Asadi Bagloee. "Applications of artificial intelligence in transport: An overview." Sustainability 11, no. 1 (2019): 189.
40. Aretakis, N.; Roumeliotis, I.; Alexiou, A.; Romesis, C.; Mathioudakis, K. Turbofan Engine Health Assessment from Flight Data. J. Eng. Gas Turbines Power 2014, 137, 041203
41. Oza, N.; Castle, J.P.; Stutz, J. Classification of aeronautics system health and safety documents. IEEE Trans. Syst. Man Cybern. Part C Appl. Rev. 2009, 39, 670–680
42. Laurell, C.; Sandström, C. The sharing economy in social media: Analysing tensions between market and non-market logics. Technol. Forecast. Soc. Chang. 2017, 125, 58–65.
43. L. Gaur, A. Afaq, G. Singh, and Y. K. Dwivedi, "Role of artificial intelligence and robotics to foster the touchless travel during a pandemic: a review and research agenda," International Journal of Contemporary Hospitality Management, vol. 33, no. 11, pp. 4079–4098, Jan. 2021, doi: https://doi.org/10.1108/IJCHM-11-2020-1246.
44. Firnkorn, J.; Müller, M. What will be the environmental effects of new free-floating car-sharing systems? The case of car2go in Ulm. Ecol. Econ. 2011, 70, 1519–1528.
45. A. Afaq and L. Gaur, "The Rise of Robots to Help Combat Covid-19," in 2021 International Conference on Technological Advancements and Innovations (ICTAI), 2021, pp. 69–74. doi: https://doi.org/10.1109/ICTAI53825.2021.9673256.
46. Raymond, R.; Sugiura, T.; Tsubouchi, K. Location recommendation based on location history and Spatio-temporal correlations for an on-demand bus system. In Proceedings of the 19th ACM SIGSPATIAL International Conference on Advances in Geographic Information Systems, Chicago, IL, USA, 1–4 November 2011; p. 377.
47. Mukai, J.; Watanabe, N.; Feng, T. Route Optimization Using Q-Learning for On-Demand Bus Systems. In Knowledge-Based and Intelligent Information and Engineering Systems; Springer: Berlin/Heidelberg, Germany, 2008; pp. 567–574.
48. Ramakrishnan, R., & Gaur, L. (2019). Internet of things: approach and applicability in manufacturing. CRC Press.
49. Chien, S.I.-J.; Ding, Y.; Wei, C. Dynamic Bus Arrival Time Prediction with Artificial Neural Networks. J. Transp. Eng. 2002, 128, 429–438.
50. Jeong, R.; Rilett, R. Bus arrival time prediction using artificial neural network model. In Proceedings of the 7th International IEEE Conference on Intelligent Transportation Systems, Washington, DC, USA, 3–6 October 2004; pp. 988–993.
51. Ziyan, Chen, and Liu Shiguo. "China's self-driving car legislation study." Computer Law & Security Review 41 (2021): 105555.

52. Huiling, B.; Goh, E. AI, Robotics and Mobility as a Service: The Case of Singapore. Field Actions Sci. Rep. J. Field Actions Spec. Issue 2017, 26–29.
53. Lim, Hazel Si Min, and Araz Taeihagh. "Autonomous vehicles for smart and sustainable cities: An in-depth exploration of privacy and cybersecurity implications." *Energies* 11, no. 5 (2018): 1062.
54. Abduljabbar, Rusul, Hussein Dia, Sohani Liyanage, and Saeed Asadi Bagloee. "Applications of artificial intelligence in transport: An overview." *Sustainability* 11, no. 1 (2019): 189.
55. Oberoi, S., Kumar, S., Sharma, R. K., & Gaur, L. (2022). Determinants of artificial intelligence systems and its impact on the performance of accounting firms doi:https://doi.org/10.1007/978-981-16-2354-7_38
56. Ma, J.; Yang, Y.; Guan, W.; Wang, F.; Liu, T.; Tu, W.; Song, C. Large-scale demand driven design of a customised bus network: A methodological framework and Beijing case study. J. Adv. Transp. 2017, 2017, 3865701.
57. Zhou, C.; Dai, P.; Li, R. The passenger demand prediction model on bus networks. In Proceedings of the 2013 IEEE 13th International Conference on Data Mining Workshops, Dallas, TX, USA, 7–10 December 2013; pp. 1069–1076.
58. Afaq, A., & Gaur, L. (2021). The rise of robots to help combat covid-19. Paper presented at the Proceedings of International Conference on Technological Advancements and Innovations, ICTAI 2021, 69–74. doi:https://doi.org/10.1109/ICTAI53825.2021.9673256
59. Chowdhury, Mashrur, and Adel W. Sadek. "Advantages and limitations of artificial intelligence." Artificial intelligence applications to critical transportation issues 6, no. 3 (2012): 360–375.
60. K. C. Santosh and L. Gaur, "Introduction to AI in Public Health," in Artificial Intelligence and Machine Learning in Public Healthcare, Springer, 2021, pp. 1–10.
61. G. Singh, B. Kumar, L. Gaur, and A. Tyagi, "Comparison between Multinomial and Bernoulli Naïve Bayes for Text Classification," in 2019 International Conference on Automation, Computational and Technology Management (ICACTM), 2019, pp. 593–596. doi: https://doi.org/10.1109/ICACTM.2019.8776800.
62. L. Gaur et al., "Capitalising on big data and revolutionary 5G technology: Extracting and visualising ratings and reviews of global chain hotels," Computers & Electrical Engineering, vol. 95, p. 107374, 2021, doi:https://doi.org/10.1016/j.compeleceng.2021.107374.
63. J. Rana, L. Gaur, G. Singh, U. Awan, and M. I. Rasheed, "Reinforcing customer journey through artificial intelligence: a review and research agenda," International Journal of Emerging Markets, vol. ahead-of-print, no. ahead-of-print, Jan. 2021, doi: https://doi.org/10.1108/IJOEM-08-2021-1214.
64. Gaur, L., & Ramakrishnan, R. (2019). Developing internet of things maturity model (IoT-MM) for manufacturing. International Journal of Innovative Technology and Exploring Engineering, 9(1), 2473–2479. doi:https://doi.org/10.35940/ijitee.A4168.119119

Chapter 2
Intelligent Transportation Technology Enablers

2.1 Introduction

With the quick improvement of urbanisation, the rate of populace dwelling in urban communities is imagined to ascend to 70% in 2050. As an effective method of transportation in urban areas, driving by vehicles is a fundamental part of everyday existence. To this end, vehicular organisations (VANETs) have drawn in a parcel of consideration to create a clever transportation framework and make the excursion more helpful and agreeable. In ITS, vehicles and side of the road, working units are remotely associated by utilising vehicular correspondence innovations to give drivers continuous traffic data and help them settle on driving choices [2]. Despite the endeavours on ITS, with vehicles constrained by person, ITS is just a data frame with the following issues:

Controlled by the driver, the driving activities of a vehicle are chosen by the driver's individual experience and emotional awareness. Also, the drivers in an exhausting ambitious climate might be diverted by different things (e.g., telephone ringing, view, and so on) along the way. The Association for Safe International Street Travel detailed that the explanation that causes mishaps is primarily due to drivers' unseemly tasks. Separated from the well-being issue, as displayed in Fig. 2.1, drivers need to do a part of work serious and tedious works (e.g., park, charge and fix their vehicles) in the ITS and consequently neglect to acquire agreeable travel insight.

The ITS can provide data more prescribe driving choices to drivers. In any case, not every one of the drivers operates as per the suggested choices. Moreover, as a general rule, the street conditions given by the ITS are the occasions that have occurred. This peculiarity predominantly comes from how the traffic status is impacted by the choices of the multitude of drivers. Therefore, it is hard to plan and make due to traffic worldwide [1, 2].

L. Gaur, B. M. Sahoo, *Explainable Artificial Intelligence for Intelligent Transportation Systems*, https://doi.org/10.1007/978-3-031-09644-0_2

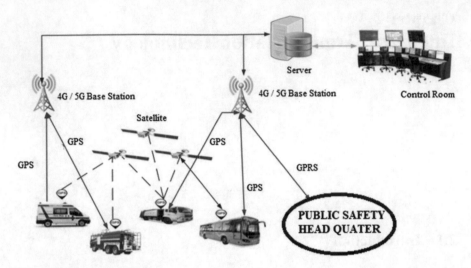

Fig. 2.1 Architecture of ITS

With different travel aims, various drivers have distinctive driving prerequisites. For instance, a few drivers might focus on the driving time, while additional drivers might often think about the driving expenses. The ITS to every driver can suggest driving is restricted, where the assorted driving necessities of drivers cannot be fulfilled. Moreover, the recommended choices in the ITS are not suitable because of the absence of exact administration and worldwide control. As a result, the drivers might experience the ill effects of bad quality of involvement (QoE) because the ITS neglects to meet their assorted necessities [3]. Luckily, the quick advancement of independent driving innovation makes it feasible to change the ITS. It is announced that all Tesla vehicles transported in 2019 are outfitted with total independent driving capacity. Inserted with different sensors, registering gadgets, and correspondence units, the controllable aerial vehicles (AVs) can comprehend the encompassing traffic climate and make clear driving choices without anyone else. This way, drivers are delivered from getting into exhausting and dangerous works, where the ongoing information discernment and precise vehicle control can essentially work on the driving well-being. Besides, by coordinating VANETs with automated reasoning, the AVs can be cooperatively made to accomplish the worldwide planning framework and fulfil the assorted driving necessities. With these promising benefits, independent driving will alter the activity method of the customary transportation framework, precisely, the ITS [4].

The chapter discusses the fundamental advances toward cutting-edge Intelligent Transportation Systems by considering accessible vehicular connectivity developments, artificial intelligence, and the parts of AVs. We first encourage the construction of independent vehicular organisations based on the rapid improvement of AVs and the development of high-level vehicular correspondence innovations to aid in identifying global information discernment and disseminating continuous information [5]. Then, at that point, we coordinate the AVNs and Simulated intelligence to plan the FITS, which is a precise and effective worldwide traffic planning and the executives' framework. From that point onward, an instance of redone way arranging is examined to assess the execution of the proposed structure. At last, we discuss the exploration issues to feature the future examination headings [6].

2.2 Autonomous Vehicles

To see the ambitious climate and make driving choices like people, as displayed in an AV, has the accompanying parts typically:

Detecting gadgets: Different sensors and devices, such as lidar, radar sensors, camcorders, the global positioning system (GPS), and an inertial estimation unit, are equipped with AVs, considered a robot. Every AV can grasp the traffic environment and achieve regulated driving with the help of various sensors and devices.

Capacity gadget: An onboard unit (OBU) is inserted to reserve significant data in every AV. Otherhand, the private data of the AV can be secured in the OBU, the interests of the proprietor. Then again, the valuable data, for example, high accuracy maps, can be committed to working with independent driving [7].

Handling gadget: The focal handling unit can coordinate the information detected from encompassing climate and the valuable information reserved in OBU to make driving choices.

Control gadget: The logic control unit's (LCU) responsibility is to complete continuing control of the locally accessible equipment gadgets to achieve securely autonomous driving. With the help of these components, the independent driving course of an AV may be divided into the associated advancements.

Climate discernment: The AV utilises a gathering of sensors to detect the encompassing static articles (e.g., street signs and the number of paths) and dynamic items (e.g., people on foot, creatures, and other moving AVs), where the gathered information are utilised to help driving choices.

Conduct choice: Behaviour choice is the interaction in which the CPU settles on driving choices and produces the orders to control the AV's movement by dissecting the apparent traffic climate data and the AV's present driving state.

Movement control: According to the driving choices made by the CPU, the LCU then, at that point, sends the comparing control orders to motors, brakes, and haggles forward through wired transmissions to acknowledge safe and exact movement control [8].

2.3   HetVNETs

Different communication technologies may be used to produce heterogeneous vehicular networks due to the development of space-air-ground integrated networks. Satellite communication: Because satellite communication has worldwide communication coverage, it may offer network services for AVs that are outside of the scope of existing vehicular communication infrastructures, such as highways. It is intended to deliver network services to AVs via specialised short-range communication. A collection of roadside units may be placed along roadways so that each AV can connect with its adjacent RSU to acquire the information it needs [9].

Cellular Networks   Cellular networks are the long-term development of 4G and 5G networks. The Third Generation Partnership Project called to enable autonomous vehicles with vehicle-to-everything services. AVs' network access services can be obtained by connecting them to cellular networks through cellular base stations.

Unmanned Aerial Vehicles (UAVs)   Unmanned aerial vehicles, equipped with dedicated sensors and communication devices, can perform various vehicular services to facilitate autonomous driving. UAVs may be used as flying vehicular nodes to improve the communication performance of RSUs and OBUs because of their adaptability to changing environments [10].

2.4   AVNS

AVNs are designed based on the characteristics of AVs and the HetVNETs. Layers AV, Edge, and Cloud comprise the network's architectural structure.

AV layer: comprises a multitude of AVs. Prepared with cutting-edge sensors and OBU, every AV can detect the encompassing climate, reserve essential data, and execute registering undertakings for independent driving. Also, the seen data can be divided between a gathering of AVs to settle on the driving choices precisely. Notwithstanding, caused by the discernment scope of sensors, AVs' choices can determine driving options dependent on restricted data.

Edge layer: is made up of edge foundations that include storage, communication, and figuring capabilities, such as RSUs, CBSs, and UAVs, among other features. In contrast to the RSU and CBS fixed hubs. The devices sent at the edge layer are closer to the AVs than those delivered at the cloud layer. Thus, the edge layer may maintain administrations nearby while exhibiting minimal dormancy. Alternatively, the edge devices may pre-process the information that has to be conveyed to the cloud layer, reducing the amount of information traffic in the cloud [11].

Cloud layer: More remarkable than the AV and edge layers, the cloud layer has more calculating capacity than any other two levels. It is possible to reserve space in this layer for verifiable information associated with the traffic on the one hand and for unverifiable details related to the traffic. It is possible, for example, to save the data of an antivirus client in a cloud server to examine the client's propensity to migrate in addition to the data previously stored. In contrast, since the cloud layer has a global view of the traffic, it can collect the data generated by the unique traffic and the AVs to make deliberate decisions and deal with the traffic, and it can do so quickly.

Features of AVNs Enabled Transportation

This part will look at the aspects of AVNs-enabled transportation to prepare for the FITS development.

Worldwide information insight:

With the high-level sensors and gadgets, the traffic climate around an AV can be productively gathered. Moreover, with the help of AVNs, the information created by the AVs and information produced by the traffic. The frameworks sent at the edge layer can gather the street areas to gain traffic insight worldwide.

Continuous information transmission:

By using AVNs, the organisation's involvement in the city can be recognised. The traffic information can be distributed amongst all specialised gadgets, which is more foundational in its approach. An AV can associate with the close-by AVs to convey notice data or offer driving data utilising V2V correspondence. Then again, the AVNs of different correspondence innovations can speak with AVs to get continuous information sharing.

Multi-facet information registering:

In the AVNs, every three layers have a processing capacity and can be used to do various jobs. For the AV layer, every AV can register tasks without help from anyone else or execute the task with different AVs cooperatively. Furthermore, the devices at the edge layer can perform processing tasks on a local level. If an antivirus program encounters a job that cannot be completed without the assistance of a third party, the antivirus program may delegate the work to nearby edge devices. Even though AVs and edge devices may assist the transportation with figuring administrations, their resources are insufficient to deal with the massive processing demands imposed by the transport. The cloud server should finish a few undertakings to provide rich assets [11, 12].

Circulated information storing:

In the AVNs, various types and quality information are kept in multiple gadgets. For instance, information about antivirus products and their owners may be stored in the cloud server to facilitate the protection and demonstration program development. The data associated with a geographic region may be sent via the gadgets located at the edge layer. For example, high-accuracy guides for the city's relatively large number of streets may be kept in the cloud server. A portion of them is designated for use by edge devices and AVs. Additionally, the information about individual traffic is appropriated in various devices to assist with continuous traffic planning. The recovery season of the stored data can be decreased as some of them can be reserved edge of the organisation.

2.5   FIT

As indicated by the highlights of the AVNs empowered transportation, in this part, as the engineering, the most common way of joining AVNs and simulated intelligence. With the AVNs, the information produced by the transportation (i.e., actual organisation) can be proficiently gathered, shared, figured, and stored. Then, at that point, because of the collected information, a virtual organisation can be created by developing the insightful advanced twin of the actual organisation. The data can be handled, investigated, and examined in the virtual network to support different vehicular applications by utilising planned AI-based calculations. With the AI-based estimates, the traffic board and booking systems and plans can work with the transportation framework in the physical network. In light of the mix of AVNs and AI, we then, at that point, plan the FITS, which incorporates the accompanying frameworks [11, 13].

Data Assortment Framework
The data assortment framework can deal with all the information and data related to transportation, including information age, detecting, and assortment. This information can be gathered to accomplish various objectives. With the investigation of the data by utilising Abased calculations [14],

1. The information about each AV and its owner saved on the cloud server may be used to develop planning strategies. For example, AI may discriminate between AVs that cause problems and those that do not.
2. The information about the components produced on each AV, such as the lifespan of various onboard sensors, may be used to ensure the AV's health and promote protection and demonstration plans.
3. To enable safe and competent autonomous driving, the ecological data collected by each AV might be employed to empower it.
4. The information generated by the traffic on each street fragment to estimate and schedule the movement of the traffic stream. Furthermore, when the crucial information is assessed in the context of the high reproduction cost in the existing framework, it may be used to replicate and test novel computations and train learning models. The transportation framework then has a global view of the traffic due to the data that has been obtained. The cloud layer may use the available assets in the FITS, such as charging stations and parking garages, by

employing global data to build AI computations and then utilising those calculations.

AV Control Framework

The AV control framework alludes to an AV's complete programmed control cycle. In this interaction, each AV utilises AI-based calculations to settle on driving choices given the gathered information from the general climate. Furthermore, the data got from the adjoining AVs or edge gadgets. With these choices, there is no requirement for human intercession in driving and halting tasks as the movement of every AV can be controlled precisely and proficiently. For the model, profound learning can work on the AV's capacity to recognise and order the encompassing articles, accordingly working with the combination of the information collected by various sensors. With the AV control framework, the precise movement control of every AV can liberate drivers from getting into exhausting driving and fundamentally decrease the likelihood of mishaps.

Intelligent Driving Framework

The cloud server's responsibility is to implement the intelligent driving framework based on information acquired from AVs and edge devices in AVNs. The cloud layer can figure out the city traffic using AI-based computations to dissect the information obtained. The cloud layer may gradually determine a driving velocity distributed to the edge devices along the specified street segment. Because of the assigned driving pace, the edge devices control the autonomous vehicles (AVs) driving on this portion of the road. The AVs participating in this roadway segment maintain a driving velocity comparable to other vehicles. However, contrary to what may be predicted, when an autonomous vehicle (AV) changes its course, its needs to adjust its driving rate are the same as the speed specified in the new street part. The possibility of congestion in each street section may be reduced, without a doubt, with the installation of traffic signals in each street segment [15].

Intelligent Traffic Planning Framework

The traffic in every street segment may be efficiently managed with the help of an intelligent driving system. After that, we'll demonstrate the intelligent traffic booking architecture, which will allow us to arrange traffic in a global way going forward. Using AI-based calculations, the cloud layer can predict the traffic flow of each street segment and then manage how the traffic stream moves along each street to break up the architecture of streets and make more adaptive use of street assets. Because of the effective design, traffic lights will not be required to supervise traffic at crossing places, for example. Additionally, rewards or compensation may be utilised to direct traffic flow differently. According to simulated intelligence-based information analysis results in the cloud layer, the framework can award or compensate different street segments. For example, using this framework, charging high prices on crowded roadway regions is possible while awarding prizes on non-clogged street segments. As a result, a few AVs will be attracted by payment and drive to non-clogged street locations, regardless of whether the non-clogged street sections need a longer driving distance. Following the precise management of traffic in different roadway segments and intersections, the framework may also encourage continuous designs to deal with emergencies in the future. For example, to ensure that salvage AVs (such as squad cars and ambulances) drive in a high-speed and obstruction-free environment, the framework will provide the best driving course and inform them of any street fragments associated with the chosen route driving course. With traffic planning, transportation becomes more intelligent, and the chance of congestion may be reduced, while the overall effectiveness of traffic can be significantly enhanced.

Electronic Exchanging Framework

To reduce the need for human assistance, it is necessary to develop an electronic trading framework that allows AVs to charge and pay fees following their practices as outlined in the FITS. With the electronic exchanging framework, any person who intends to participate in the FITS is provided with a virtual currency account (VCA), which allows them to complete exchanges with various kinds of money. Every AV owner must guarantee that their AV has enough currency in its VCA ahead of time to ensure that their AV can interact with various devices efficiently. It is also created for those who do not have access to AVs so that they may travel via public transit if they so want.

Following the completion of the exchange, the associated expenditures will be organically distributed among all members, decreasing the number of AVs that need to be stopped and avoiding the possibility of an inappropriate instalment caused by a human error. Because the assets in city transportation, including parking places and charge stations, are limited and continually developing, artificial intelligence-based computations may be used to value these assets gradually. When an AV charges, consumes a parking place or executes other usage applications, the associated expenditures will be taken from the proprietor's VCA following the AV's ID, which will be deducted from the proprietor's VCA.

Individuals who go by public transportation might benefit from AI-based computations to better arrange their biological characteristics to pay for administration more advantageously. Due to this arrangement, individuals in the FITS will no longer be required to pay for exchanges using cash or credit cards. It can significantly enhance persons' mobility experiences while also accelerating the development of AVs' usage pace.

Tweaked Administration Framework

In the wholly controlled rush hour gridlock climate, travellers do not have to zero in on the driving cycle and focus harder on the driving necessities. For the most part, various individuals have diverse driving prerequisites (e.g., driving time and expenses) in any event for the equivalent trip. Hence, we also need to build up the redid administration framework in the FITS to fulfil various clients' requirements. For instance, when an AV plans to stop with minimal expense while could not care less with regards to the driving distance, the modified administration framework then, at that point, gives the ideal decision to the AV for stopping utilising the AI-based calculations.

In traffic dispatching, the FITS can charge/repay the expense to diminish/increment the movement season of clients. The modified help framework can set specific charging/pay plans for clients with various driving prerequisites by planning AI-based calculations to examine the gathered information. In particular, if a client intends to reduce movement time by choosing the briefest way and the quickest speed, it needs to pay the framework for the driving solicitation. The comparing remuneration will be delivered to the clients who will build their movement time. Also, as indicated by clients' prerequisites, Artificial intelligence-based calculations likewise can be investigated to tweak their driving courses. The tweaked administration framework can grease up the city traffic planning and upgrade the movement experience of every client.

2.6   Digital Maps

As we searched for many fundamental advances that track down specific applications all through the field of ITS, we set up that one such innovation is computerised maps. In this circumstance, computerised maps track down the application in explorer data frameworks, traffic the executives' frameworks, business vehicle activities, public transportation frameworks, and rustic transportation frameworks [1, 15].

The guide explanation is a graphical representation of the standard and manufactured items surrounding us. It offers a "whiz" method for visualising situational data and the observer's rapid absorption of that data, such as traffic thickness, speed, and travel routes. Many other graphical presentations do not achieve the data thickness possible with maps; a normal USGS geographical quadrangle is estimated to hold 250,000 bits of data per square inch, essential information.

A well-known proverb is that "words typically cannot do a picture justice." The use of guides may be traced back hundreds of years. Their availability is credited with the exploration of the globe by the Portuguese, the Spanish, and other European nations throughout the fifteenth, sixteenth, and seventeenth centuries, among other times. Early guidelines were drawn on mud tablets and, in some cases, etched in stone, as in the case of the Yii Chi Thu of the eleventh century AD. Papyrus, vellum, material, and, most recently, paper were used to create them as they became smoother and lighter in weight as time progressed. With the introduction of high-end personal computers (PCs), the creation of digital maps became possible.

The substance of a computerised guide can be arranged into two major information types:

1. realistic information
2. nongraphic information.

Practical information is advanced portrayals of guide highlights used to re-make the guide on a presentation gadget or paper. Six components commonly portray map highlights, mainly focuses, lines, regions, framework cells, pixels, and images.

Then, Nongraphic information illustrates the attributes, characteristics, or connections of guide highlights and geographic areas, regularly alluded to as element ascribes.

Depending on the application, practical information is stored either as vector or raster information. A vector representation of geographic highlights consists of focuses, lines, or polygons organised according to x and y-axis progression. Focuses may be used to address building areas, lines could manage streets or boundaries, and polygons could be used to address states or provinces. They use a topological technique to capture and regulate the meaningful links between guide highlights and geographic information.

Using this practice, line end focuses and convergences are marked on a diagram by placing hubs or vertices at the lines' points to finish and meet. The lines on the figure refer to circular segments or links shown. The bends might be uncoordinated or well-coordinated. Inferred by an undirected bend, growth between two hubs may occur in either the left or right direction. A coordinated curve indicates a bearing of movement or stream, e.g., a single-direction road.

A grouping of bends and hubs interfacing any two hubs in the chart is a chain. Like a political region, shut regions limited by a chain are addressed by polygons. Steering calculations utilise topological data to choose vehicle courses.

This equivalent data also upholds proficient polygon overlaying techniques used for spatial investigation. The clear credits of the vector information highlights are put away in an element characteristic table. The information is typically accessible in inclusions. An inclusion is an advanced portrayal of a solitary layer in a guide, like roads, structures, or districts.

Cartesian and scope/longitude are the two direction frames most often used in navigation. The values of the coordinates are presented via a variety of possible guide projections. The State Plane Coordinates (SPC) and the Universal Transverse Mercator (UTM) projections are the most often used in the United States of America. In addition, the scope/longitude coordinate framework maintains exact estimates relative to the equator and the Greenwich Prime Meridian, and it is used wherever in the globe. Most information providers deliver information in this framework; the translation of the information into other projections is then passed on to the customer. Raster information divides a geographic area into a consistent, rectangular network of lines and sections, known as a raster network. A cell is a term used to refer to the point at which a line and a section meet. Each cell corresponds to a geographic location on the Earth's outermost layer [1].

The aim of the raster information depends on the size of the cells; thus, it is critical to choose the optimal lattice size to meet the application's requirements. Instead of assigning ascribes to natural elements, as is the case with the vector structure, specific information in the raster information structure is given to the individual cells. There are two types of raster information: lattices and images. Every cell is assigned a quality value based on network information. Every lattice point sends a value that describes the Earth's surface in that particular location because of network information. Advanced landscape models (DTM) and computerised height models are two types of rising information often stored in lattices (DEM).

The cells are referred to as pixels since they contain visual information. The size of a pixel is often determined by the imaging technology used to gather the information. The source of information might be either symbolism obtained by a satellite and amended for movement, optics, and projection, or symbolism carefully inspected from a printed map, depending on the circumstances. Compared to raster information, vector information uses far less PC mass memory for storage capacity in general. Recent improvements in realistic presentation technology generate more interest in ITS applications than in raster information, which is positive.

Database Management

One superior quality of practical information is their huge memory necessities, which request proficient hunt and recovery calculations. To deal with incredibly enormous information, use is made of particular information structures. The information stockpiling models utilised is a progressive, organisation, and social. These models portray the practical design of a data set using trees, assortments of hubs and connections, and tables. They all present the client's perspective on a pattern as far as a record structure.

The various levelled information model is the most seasoned data set models. The model stores information in a structure that depends on coordinated and one-to-numerous connections; i.e., a progressive data set comprises timberland of trees. Each case of a record type can have one parent record. The advanced information structure makes it appealing; in any case, joining this straightforwardness with the culmination of the relations, for the most part, required ends up being lumbering. The outcome is that the data set item will be mind-boggling and prohibitive. An eminent reality about the progressive model is that there is no normal legacy for various levelled frameworks.

This absence of shared traits has forestalled the improvement of guidelines; the ramifications are that clients of one various levelled framework have no relocation way to another progressive framework. In the organisation model, records of a similar sort are assembled in applied documents that model balanced, one-to-many, and many-to-numerous connections. Depending on the record type, there may be a few parent record types, and depending on the record example, there may be many parent record instances. The organisation model is more flexible than the progressive approach, so it is preferred.

The model is based on a design created by the CODASYL Data Base Task Group in 1971, and it was first shown in public in 1972. Compared to the others, the only distinguishing characteristic of this model is its reliance on named data sets to detect information links. Documented records are used to address components, and these records include essential attributes and information objects that have been constructed. The settings are among the record types that may manage the relationships between substances. It is allocated to one record type, the role of set kind owner, and at least one record type is assigned the role of persons [1, 31].

The set is utilised to catch connections between the proprietor and at least one individual. The social model finds its beginnings in crafted by Codd. It offers reasonable effortlessness, upheld by a solid hypothetical establishment. At its center is the hypothesis of information conditions. The model's objective is to decide the number of relations that ought to be in the data set and the virtual credits.

The social model can be considered a grid design of lines and sections. The lines compare to data sets, whereas the areas correspond to credits. Each tuple corresponds to a logical representation of a particular information item. Connection plots consist of a collection of characteristics, while relations are made up of a group of tuples with the same plan. During the decades after its introduction in 1970, the social model has evolved into a model of fundamental syntactic components with impressive semantic capabilities.

Spatial Analysis Provides the breakdown information contained in the computerised map. Spatial connections can be categorised as one of two classifications: straightforward or topological. An illustration of a direct spatial connection would be the location succession along one side of a road section; e.g., 203 Broadway Avenue falls somewhere between 201 and 205 Broadway Avenue. Such a nearness connection was not settled from the information given in the road data set to the Broadway Avenue Road portion. Then again, topological connections are more intricate and regularly allude to the network, nearness, closeness of focuses, and polygons.

A link is vital for the polygon that describes an area of interest; availability algorithms would expose all of the connections that share that hub, regardless of whether or not a specific reference is relevant for the ties inside a polygon.

For example, in the case of contiguousness calculations, they would reveal the layout of locations that comprise a particular district. For example, proximity calculations might show the arrangement of foci adjacent to a line being studied. For example, what public scales are within a 5-kilometre radius of the freight pick-up location would be a nearness question. Spatial examination instruments that do these functions are often included with topographical data framework programming bundles such as MapInfo and ArcView. Despite the above, spatial examination computations are used to build realistic overlays and to address coordinating or geocoding, among other things.

A realistic overlay activity consolidates the highlights and relevant remarks from several guide layers into a single action that looks realistic. Consider the following scenario: a client layer is stacked on a zip code restriction layer. A satellite image of the St. Louis metropolitan area is superimposed on top of the image. After that, the watcher would have access to information regarding the geographic dispersion of its customers, which might be utilised to design practical assistance conveyance courses.

Address coordinating calculations present the capacity to produce areas that address highlights or occasions on a guide containing a road organisation. A coordinating estimate gauges a location area on a directory by introducing the given location between two road fragments utilising the location range relegated to every road section. Even though geocoding does not give the specific location area, it ordinarily offers an above and beyond response. Geocoding is because of the suspicion that:

1. addresses are mathematically consecutive, and
2. Even odd addresses happen on inverse sides of a road.

The irregularities indeed do exist, which can prompt coordinating issues. The interaction does not work in different nations, a good example being Japan, where the absence of tending to conspire makes such a location matching procedure inconceivable.

Before computerised maps, mapmakers made printed copy maps, their essential specialised objective being to make profoundly detailed guides. Since the present advanced principles are something like a developmental advance in map making, the requirement for exactness remains [1, 31].

The positional precision is the most generally involved measure of the realistic elements on a computerised map. Positional precision estimates the difference in the place of a map highlight from its actual position, and the estimation can be either relative or outright. Comparable positional exactness evaluates a component's position precision on a guide compared with the status of different highlights in a similar direction.

Outright positional precision estimates the position exactness on a guide comparative with its actual situation on the Earth's outer layer. The guide scale is vital to evaluating positional equivalence. The ratio of the guide's estimate units to the guide's estimation units on Earth is a common way to convey the size. For example, a measurement of 1 centimetre in the direction corresponds to 100,000 centimetres on the surface of the Earth. The magnitude of the positional error rises in direct proportion to the scale.

The range is likewise a significant component in the presentation of guides. As the map scale decreases, the thickness of information per region unit becomes more noteworthy. As massive changes in scale happen, it is essential to eliminate detail to deliver an outwardly satisfying guide. The opposite circumstance is likewise obvious; a critical expansion in scale requires added detail. One more significant proportion of a computerised map's realistic exactness is fulfilment, particularly valid for ITS applications in-vehicle route and course arranging.

Two kinds of guide fulfilment blunders is recognised in writing:

1. calculation mistakes, and
2. geography mistakes.

Calculation blunders connect with the positional precision of the street information on the map. Their essential effect is on the proper position on a guide utilising scope; longitude arranges as dictated by a few programmed vehicle area strategies, e.g., GPS retribution. When the mistake is enormous, the vehicle is situated off the street.

This mistake is basic to applications that utilise raster road maps. It is impossible to partner the vehicle area to the road it is voyaging shows the effect of mathematical mistakes on a vehicle route framework. Geography mistakes allude to blunders in the network of the road fragment information in the guide, and they incorporate both missing and no existing roads.

In the space of vehicle route, this mistake prompts incorrect guide matching where such a strategy is utilised for a map route. Geography mistakes most considerably affect vehicle steering calculations since they lead to the age of courses over imaginary streets. It impacts the acknowledgement of vehicle route frameworks in a way that may be messed up regarding the degree of mistake event, yet which focuses on the significance of utilising something like date street information bases.

A critical wellspring of topological mistakes is utilising development manufacturers' plot plans in computerised maps; as a rule, all initially arranged roads were rarely evolved. Road data sets that depend exclusively on TIGER information documents will often experience the ill effects of topological blunders.

Philosophy and patterns for a keen vehicle framework in developing nations: The fast development of the total populace represents an issue of mobility for street clients. Vehicle traffic includes controlling traffic in street organisations, decreasing traffic accidents, and developing traffic streams, particularly the judicious utilisation of road resources. The expressway code, need rules, traffic signs, traffic lights, and cops' traffic have been added to traffic guidelines. Innovative advances in telecommunication have prompted the introduction of the clever vehicle framework. There are mechanised frameworks that offer multiple applications to move frameworks dependent on refined telecommunication foundations like sensors, cameras, and cell phones. Likewise, these frameworks provide a few benefits [1]: increasing road security, assessing ecological contamination, and determining traffic flow and spread data. These frameworks incorporate vehicular ad hoc networks that allow correspondence between vehicles, and other foundations like Road Side Unit (RSU). RSU is the framework that is sent out for the improvement of correspondence between vehicles. They also increase the general inclusion of a vehicular organisation. Several developed nations utilise such frameworks. Should not something be said about implementing such frameworks in creating countries? Transportation frameworks in emerging nations are confronted with a few fundamental factors. We lack value street infrastructure: the more significant part of these nations does not have excellent street infrastructures. It implies that any vehicle goes on similar tracks. The lack of media transmission foundations is another issue in these nations. The foundations there do not permit inadequate sending of correspondence between substances that constitute the street transport framework. We additionally have the shortfall of an ineffective framework for observing exercises out and about the network.

Ridesharing with a smartphone: Ridesharing is a new concept of transportation that depends on data sharing between individual explorers to function well. An expedition with individuals with similar agendas and schedules includes sharing a car and pooling travel expenditures such as petrol, rest stops, and toll charges. With the evolution of personal digital assistants (PDAs), multiple programs have been developed to function on various ridesharing services. Among the many benefits is the reduction of travel time, the alleviation of congestion, the reduction of fuel use, and the reduction of air pollution. Advanced mobile phones, such as Apple's iPhone software and Google's Android platform, provide real-time ridesharing systems that are simple and very appealing, such as the iPhone software and Google's Android platform. Modern cell-based ridesharing is often accomplished via the use of many apps. The ridesharing apps are divided into two categories according to their genre partner.

First and foremost, a program that provides ridesharing as an optional toon-request taxi administration is prohibited. Toxify and Uber are examples of apps under this transportation services category. Furthermore, usage categorisation adheres to the traditional characteristics of ridesharing services such as BlaBlaCar, Hitch-A-Ride, Via, and Split.

A model for road traffic control is as follows: It is found that there were deficiencies in transport frameworks in developing countries, such as the absence of expressways, the absence of telecommunications infrastructure to foster intelligent vehicle frameworks, the need for management arranging frameworks in transportation frameworks, and the lack of public vehicle administrative administrations. In light of these inadequacies, it is understandable why it is challenging to implement effective transportation networks in developing countries. On the other hand, in Cameroon, the gendarmerie uses drifting vehicle information to limit the speed of vehicles on specified sections of the road. Cameroon Customs also uses GPS to track the progress of product trucks in the street organisation and other applications.

The Cameroonian vehicle structure undoubtedly allows them to fulfil their goals, mainly when dispatching a disproportionately large number of trucks on the country's road network. They enable the flow to continue without interfering with the high level of management provided by the street organisation. These inquiries eventually led to a discussion regarding the modes of transportation used in these nations. Roads are shared by various modes of transportation, including public transportation, trucks, logging trucks, private automobiles, and rustic vehicles (transport action hefted around huge urban areas to the surrounding villages involving vehicles in vile states). The revolt in the mode of transportation witnessed in these countries justifies the high rate of accidents in the segments that separate the metropolitan regions of these countries. A co-organising physical and data frameworks approach is proposed to normalise that insurgency inside the vehicle framework. Focusing on the social, and financial aspects that exist in these nations presents a strategy for standardising the vehicle frameworks in these countries. Despite this, the problem of guidelines continues to be a concern.

Developing a traffic observation framework in an interurban street transport network requires a model whose primary goals are to maximise the fulfilment of street clients. Also, allowing them to reduce their venture times, characterise an emotionally supportive network that can manage all of the aggravations observed during the time spent regulating metropolitan street traffic, and develop a communication system.

The first phase of this effort consists in creating a graphical representation of the problem that needs to be addressed. The framework's parameters, destinations, and imperatives are considered while deciding on a visual representation. The metropolitan transport network model serves as a guideline model not so much for a network component as it does for a crossing point; instead, it serves as a guideline model for a metropolitan condition. A few models of depictions of urban automobile networks in the literature are worth mentioning. The following are examples of vertices or nodes that may be examined in a street transportation organisation: towns, toll booths, gauging stations, and customs designated locations. Circular segments are highly regarded and, depending on the nature of the problem to be resolved, may effectively translate the characteristics of the features they address.

The value of an arc can be determined by the design of the organisation on the segment (sinuosity and arrangement of the tracks), the nature of the framework (number and width of ways), geographical limitations (slope), power guidelines (speed), specialised qualities of the vehicles, a halt in the system's activity (a blockage), the length of the segment, and the limit of the component (the number of vehicles allowed to travel on that segment at a given time). Alternatively, This segment's goal is to audit the authoring of particular models of urban transportation networks and to provide a report along these lines. To determine if the strategies can be used across different urban environments or whether a combination of approaches can be applied.

Traffic Management and Control
The previous research that laid the foundation for ITS has been done for over a third century. In 1960, robotised roads were introduced, followed by electronic course guidance frameworks in 1968–1970, and finally, the Comprehensive Automobile Traffic Control System (CACS) in 1979. A comparison is made between the advancements made in intelligent transportation systems in the United States, Japan, and Europe. The Advanced Traffic Management System (ATM) is the public foundation element of the ITS systems. It's the ITS foundation, and it will take in any remaining specialist ITS aspects with whatever support it can get. The ATMS innovation impacts electronic cost assortment, traffic surveillance, and versatile traffic-light control [15]. The capacities incorporated by ATMS and prominent include:

- assortment of information from vehicles to produce framework comprehensive traffic data,
- use of information about the traffic stream and the street framework and its conditions to enhance traffic activities, and
- interchanges between the street framework and the vehicles and voyagers.

Three central points have driven the coordination of ATMS endeavours in the US:

1. legislative financing for ITS exercises, alongside some particular legislative instructions,
2. ITS America, its functioning gatherings on the ATMS, its essential arrangement, and
3. the J.S. Speck direction, including the board of assets for exploration, testing, and sending of ATMS-related innovations.

Ofsted gives a fantastic conversation on this subject. In this part, we outline ATMS functional field preliminaries, philosophies for displaying and planning traffic stream, and traffic information for the executives.

An approach for showing a traffic checking system connecting urban street transportation networks has been developed in emerging nations. Compared to a few other current models in writing, our postal-friendly framework is a dispersed framework that provides adaptation and autonomy in the modules that have been offered. Problems are addressed in these courses, and some streamlining strategies are proposed. For these optimisation difficulties associated with an intelligent transportation system, we used autonomic registration, a newly emerging computing paradigm, as our approach.

Consequently, to be helpful in the creation of countries, our offer is a fourfold worldview: self-configuration, self-mending, self-improvement, and self-assurance, among other things. It is re-nationalised in the proposal via the use of four massive components. The establishment of an ITS project in developing nations enhanced the infrastructure of the transportation systems. Financially, it provides a few opportunities for long-term viability. It will give a great deal of employment for young people in nations concerned about various issues. It is undeniably true that most of them have high unemployment rates.

The development, installation, and organisation of control foci will need many workers, resulting in the creation of several vocations. Because of the decisions that have been made concerning this framework, the bodies that will be in charge of dealing with this observation framework (whether they are state or private institutions) will be prepared to accept large sums of money as a form of compensation from the said organisation. Finally, it should not be forgotten that hotels and basic shops might be set up in the vicinity of treatment control centres to provide a place for street customers to relax between treatments.

Because of this, we are considering characterising an isolated framework in which each module addresses a contributor to the problem and assumes a feasible solution following optimisation models and execution rules that we shall define later on in this paper. The ability to work on checking exercises over the metropolitan road network may be used in our working environment. Because of this technique, the framework is easy to comprehend, administer, and develop [16].

As an aftereffect of this work, street well-being remains a problematic issue in developing nations like Cameroon, where there are numerous deaths due to street mishaps, particularly in metropolitan vehicle networks. There are various explanations behind these weaknesses, including the absence of sufficient street framework; the lack of telecom Muni-cations foundations vital for the arrangement of intelligent transport frameworks as experienced in created nations.

Existing models of street traffic guidelines provide arrangements in most cases only regarding a convergence traffic signal by offering static or dynamic deals (because of improvement strategies) on the management of traffic signals and line the board. The demonstration of urban transport organisations, then again, proposes strategies to regulate traffic in urban communities [17].

However, the need associated with the setting of urban street networks in developing economies does not permit the model to be suggested only based on a single method. Our research intends to offer a system for monitoring the traffic of interurban transportation networks in developing countries reliant on information and communications technology (ICT), which may be found in the poorest of countries [18].

Based on the demonstration of urban automobile networks, this crossover architecture is organised around four essential parts. The autonomous administration of control focuses on the introductory module of this application. It has the authority to regulate circumstances involving hand-offs and oversee exercises between the cities (designated spots). A communication design between these control focuses is defined by these control focuses' presence.

It is recommended that you extend correspondence coverage while simultaneously limiting the number of control focuses on achieving optimal delivery of this control focus via the urban street network. It allows for the development of solutions for dispersing irritations in a specific region of the inter metropolitan street organisation to aid in reducing congestion between metropolitan road networks. The final module is course arranging in a dynamic environment.

It will function in conjunction with the aggravation location and dispersion module to provide customers with various cycling options while also reducing congestion on the street transportation network. Finally, the final module is concerned with the examination of the parts. It relies on independent decision assistance, based on the information given by various modules and the autonomic computing paradigm, to notify network clients about punishments associated with their actions and choices.

This model should diminish street mishaps, demonstrations of incivility by specific clients, and exercises commonly conveyed out in the street organisation. The work of defending the principal module is independent administration of the control points. It pointed to characterising a procedure to enhance the administration of control focuses between metropolitan vehicle street networks in developing countries.

Intelligent Traffic Management
As previously stated, improving the operational efficiency of the transportation system is one of the most important design goals of the Intelligent Transportation System (ITS). Developing intelligent traffic management software is now essential in achieving the goals [19]. The use of sensors and computers is the primary means of attaining intelligent traffic control at the time. A traffic signal control (TSC) system is used to execute real-time dynamic adjustment of real-time traffic flow using the real-time traffic flow approach. Also, to ensure that traffic flows as smoothly and reliably as possible within the permissible flow range of the traffic infrastructure. This strategy is characterised as a reactive method. The topic is now experiencing a research trend that focuses primarily on using reinforcement learning (RL) technology to develop the optimal signal control scheme via trial and error under the supervision of suitable reward functions. Several approaches have been proposed to deal

with this problem, including, for example, effectively managing traffic flow in the transportation system and ensuring that the system runs smoothly in all situations [20]; such techniques are essential. While the advantages of this kind of approach are evident, the disadvantages of this type of method are also apparent. When the volume of traffic exceeds the capacity of the road, this method is rendered ineffective.

While these efforts are in place, they will not be able to alleviate the traffic congestion that has formed [21]. When it comes to traffic management approaches, the proactive traffic management technique based on traffic prediction may assist the reactive method in avoiding traffic bottlenecks to a certain extent. Still, the aggressive process cannot help prevent traffic bottlenecks entirely. As a result of the use of traffic forecasting, several advantages are observed: First, it includes the traffic signal control system (which can only passively regulate current traffic flow at an intersection; based on the inherent traffic limit of the lane) rather than fundamentally regulating traffic flow; but, cannot radically restrict traffic flow. The proactive strategy based on traffic forecasts may be regarded as a macro control method for proactive traffic management. When predicting traffic flow, the system can account for the probability of congestion at the crossing by using mathematical formulas. Let us recall that the transportation system, being a complicated and large-scale system, needs both a reactive strategy for micro regulation and a proactive approach for macro regulation to work correctly. These approaches are complimentary as well as necessary. It has been proved in [22] that if there is no adequate traffic signal management technique in place, even when the traffic flow is lower than the road design capacity, traffic congestion will occur regardless of how low the traffic flow is. A traffic bottleneck that happens at a single junction has the potential to set off a chain reaction of failures that may eventually result in system-wide congestion [23].

Autonomous Driving
Because vehicles are such essential participants in the transportation system, increasing the driving safety of vehicles is vital to enhancing the overall safety of the transportation system. As a result, the desire for increased traffic safety has also fuelled the development of autonomous driving technologies in recent years. A broad division may be made between the several types of autonomous driving technologies that are now available. The first category includes driving aid technology connected to the vehicle's movement state or the driver's status, which has evolved and slowly introduced into the market. Several of these techniques, such as lane keep/departure warning systems [24], driver drowsiness detecting systems [25], and others, are classified as Level 1 autonomous driving techniques according to the SAE standard [26]. Some are classified as Level 2 autonomous driving techniques, such as dynamic vehicular route planning [27]. The second sort of approach is concerned with detecting traffic situations in the vicinity of cars. This kind of technology has as one of its essential purposes the improvement of traffic-related accurate detection technology for traffic-related objectives, such as animal detection [28], vehicle detection/model identification [29], and pedestrian detection [30], among others. The Convolutional Neural Network (CNN) is used to extract the feature information of different targets contained in a picture by utilising the massive

computing power of the GPU. It then identifies the detected target using the corresponding classification method, which is currently the most widely used (e.g., SVM used in RCNN, SoftMax used in Faster RCNN, etc.). For both systems to function correctly, they must have the same prerequisite. The necessary targets must be included in the pictures acquired by the onboard camera or laser rangefinder.

Most of the time, however, due to obstacles or the restricted viewing angle of the detection equipment or lighting circumstances impairing sight, the detecting equipment hits pedestrians. Due to the high frequency with which the targets are in the blind region of the detection vehicle, it is difficult for the detection vehicle to detect them correctly. According to statistics from the United States Department of Transportation, walking in low-visibility scenarios, such as low-illuminating conditions or being in each other's blind zones, is responsible for around 75% of pedestrian-related events, according to statistics from the United States Department of Transportation [30]. Consequently, establishing the most effective way to detect low-visible pedestrians and achieve other objectives is a big challenge that must be solved. • The following are some of the advantages that may be gained by using traffic forecasting: The use of collaborative detection, which uses the information-sharing capabilities of the ITS, as a potential solution to the fundamental issue may be considered a feasible alternative to investigate further. Another way of putting it is that, because of the discrepancy in position between the automobiles on the road and the angle at which the detector is pointed, pedestrians and other targets in the blind zone of a specific vehicle may seem utterly unhindered from the perspective of other vehicles. Using vehicle-to-vehicle communication (V2V), cars may collaboratively share information about each other's seen targets to improve the perception of items such as persons detected in their respective detection blind zones. As a result of the inclusion of traffic flow prediction, the efficiency of connected, collaborative detection activities may be increased indirectly [32–38]. The reason for this is because, as previously said, traffic forecasting has the potential to dramatically improve the efficiency and dependability of data transport in an Internet of Things context. It is also possible for vehicles to predict the anticipated traffic volume before approaching a route stretch. It is adequate in terms of vehicle density on a given road section to allow collaborative detection, which may be used to improve the detection accuracy of low-visible pedestrians and achieve other purposes. If the relevant prejudgment result does not support cooperative detection, the vehicle control system should take precautionary measures to avoid collisions, such as slowing the vehicle down. As a result, there would be a lesser risk of pedestrian injury due to collisions. Traffic forecasting may also be utilised to improve the accuracy of dynamic vehicle route planning cost-effectively and directly. With accurate traffic volume forecasts, cars may predict and avoid traffic congestion in advance, increasing their comfort and efficiency and reducing congestion on the road.

   This section discusses how incorporating traffic prediction has increased the performance of critical ITS-related applications. The following two areas of the performance of ITS are improved as a result of traffic prediction in general:

   Improved efficiency and dependability of data transmission are two essential goals. Based on the information presented above, we can conclude that the successful distribution of data inside the ITS system is the foundation for executing numerous ITS applications. Generally, a vehicle is a principal participant in data transfer and the primary contributor of processing power and communication resources in the system in which it operates. Accurate traffic prediction may considerably increase the efficiency with which data routing is established and the accuracy with which idle computing or data storage capacity is forecasted across the system. These are essential for guaranteeing the smooth functioning of associated ITS applications regularly.

   The second point is that an accurate traffic flow forecast may aid relevant transportation systems or agencies develop appropriate matching traffic management plans, enabling traffic to be more fairly dispersed across the transportation system, as previously indicated. Therefore, they decreased the possibility that large-scale congestion would arise inside the system due to heavy traffic flow on a given day or at a specific time. Effective traffic flow prediction enables vehicles to anticipate and adjust their operation habits to avoid congestion at the micro-level, improving passengers' travel comfort.

   The intelligent transportation, this study recognises AI's revolutionary potential and how it can be a paradigm-shifting force that will completely revolutionise the way people move. Autonomous, digitised, and networked transportation services are not adequate if they do not help to increase environmental conservation, resource efficiency, productivity benefits, and social inclusion and integration [39–44]. People must think that a shift is possible, actively participating in the new urban ecosystem and taking advantage of the possibilities afforded by the AI-transport-nexus as given in Table 2.1.

2.7 Conclusion

The chapter explored how traffic forecasting may enhance the performance of the associated ITS applications. The chapter discussed the overall prediction technique and some fundamental ideas of traffic flow forecasting and forecasting to provide a more thorough picture of the traffic forecasting function in intelligent transportation systems. Considering the current norms and standards for autonomous boats and the worldwide efforts in this sector, it is clear that such operations include several significant industry actors and national and international organisations and regulators working together. With the considerable growth in the number of vehicles on the road, traffic congestion in the city is becoming worse.

Table 2.1 AI-transport nexus

Artificial intelligence	In computing, AI is defined as a machine's ability to simulate the human mind by interpreting data received from its environment, learning from it, and applying it to complete tasks, even in the most unexpected and novel scenarios
Connected and autonomous vehicles (CAVs)	Intelligent and autonomous vehicles are a kind of vehicles capable of recognising and comprehending their environment and driving, navigating, and acting appropriately without the assistance of a person. At the same time, it has connectivity traits that enable it to be proactive, cooperative, informed, and coordinated in its activities
Personal aerial vehicles (PAVs)	Personal aerial vehicles are flying people movers, bridging the gap between scheduled airliners and ground transportation by using available open air space to provide new levels of quick, on-demand urban mobility
Unmanned aerial vehicles (UAVs)	Unmanned aerial vehicles (also known as drones) are intelligent aircraft that can fly without onboard pilots and provide reliable air transport solutions to improve military, law enforcement, and commercial services. Drones are becoming increasingly popular to enhance military, law enforcement, and commercial services. UAVs are becoming more popular in various uses, including military, law enforcement, and commercial
Mobility-as-a-service	Mobility-as-a-service is a system that uses an all-in-one digital platform to provide integrated travel planning, booking, smart ticketing, and real-time information services on a subscription or "prepay" basis, replacing privately owned cars
Industry 4.0	Technology such as cyber-physical systems (CPS) and the internet of things (IoT) are changing industrial processes into industry 4.0. A transformational paradigm reflects the computerisation, automation, digitisation, and information of industrial systems
Physical internet	The physical internet is a global term for digital, automated, connected, and big data technologies that make it easier for physical objects to get where they need to go and how they are used for multimodal freight transportation and logistics

References

1. Garcia-Ortiz, A., S. M. Amin, and J. R. Wootton. "Intelligent transportation systems—Enabling technologies." Mathematical and Computer Modelling 22, no. 4–7 (1995): 11–81.
2. Gaur Loveleen, Bhandari Mohan, Bhadwal Singh Shikhar, Jhanjhi Nz, Mohammad Shorfuzzaman, and Mehedi Masud. 2022. Explanation-driven HCI Model to Examine the Mini-Mental State for Alzheimer's Disease. ACM Trans. Multimedia Comput. Commun. Appl. (March 2022). doi:https://doi.org/10.1145/3527174
3. F. A. Silva, A. Boukerche, T. R. M. B. Silva, E. Cerqueira, L. B. Ruiz, and A. A. F. Loureiro, "Information-driven software-defined vehicular networks: Adapting flexible architecture to various scenarios," IEEE Vehicular Technology Magazine, vol. 14, no. 1, pp. 98–107, Mar. 2019.
4. National Highway Traffic Safety Administration (NHTSA), Critical Reasons for Crashes Investigated in the National Motor Vehicle Crash Causation Survey, 2015.
5. S. R. Rizvi, S. Zehra, and S. Olariu, "Aspire: An agent-oriented smart parking recommendation system for smart cities," IEEE Intelligent Transportation Systems Magazine, vol. 11, no. 4, pp. 48–61, Winter 2019.

6. J. Liu and J. Liu, "Intelligent and connected vehicles: Current situation, future directions, and challenges," IEEE Communications Standards Magazine, vol. 2, no. 3, pp. 59–65, Sept. 2018.
7. N. Cheng, W. Quan, W. Shi, H. Wu, Q. Ye, H. Zhou, W. Zhuang, X. Shen, and B. Bai, "A comprehensive simulation platform for space air-ground integrated network," IEEE Wireless Communications, vol. 27, no. 1, pp. 178–185, Feb. 2020.
8. A. Bhat, S. Aoki, and R. Rajkumar, "Tools and methodologies for autonomous driving systems," Proceedings of the IEEE, vol. 106, no. 9, pp. 1700–1716, Sept. 2018.
9. Z. MacHardy, A. Khan, K. Obana, and S. Iwashina, "V2x access technologies: Regulation, research, and remaining challenges," IEEE Communications Surveys & Tutorials, vol. 20, no. 3, pp. 1858–1877, Feb. 2018.
10. Y. Hui, Z. Su, T. H. Luan, and C. Li, "Reservation service: Trusted relay selection for edge computing services in vehicular networks," IEEE Journal on Selected Areas in Communications, vol. 38, no. 12, pp. 2734– 2746, Dec. 2020.
11. H. Peng, L. Liang, X. Shen, and G. Y. Li, "Vehicular communications: A network layer perspective," IEEE Transactions on Vehicular Technology, vol. 68, no. 2, pp. 1064–1078, Feb. 2019.
12. Hui, Yilong, Zhou Su, Tom H. Luan, and Nan Cheng. "Futuristic Intelligent Transportation System." *arXiv preprint arXiv:2105.09493* (2021).
13. W. Shi, H. Zhou, J. Li, W. Xu, N. Zhang, and X. Shen, "Drone assisted vehicular networks: Architecture, challenges and opportunities," IEEE Network, vol. 32, no. 3, pp. 130–137, May 2018.
14. Z. Su, Y. Hui, and T. H. Luan, "Distributed task allocation to enable collaborative autonomous driving with network softwarization," IEEE Journal on Selected Areas in Communications, vol. 36, no. 10, pp. 2175–2189, Oct. 2018.
15. Gaur Loveleen, Bhandari Mohan, Bhadwal Singh Shikhar, Jhanjhi NZ, Mohammad Shorfuzzaman, and Mehedi Masud. 2022. Explanation-driven HCI Model to Examine the Mini-Mental State for Alzheimer's Disease. ACM Trans. Multimedia Comput. Commun. Appl., March 2022. doi:https://doi.org/10.1145/3527174
16. S. Gyawali, Y. Qian, and R. Q. Hu, "Machine learning and reputation based misbehavior detection in vehicular communication networks," IEEE Transactions on Vehicular Technology, vol. 69, no. 8, pp. 8871–8885, Aug. 2020.
17. S. Zhang, J. Chen, F. Lyu, N. Cheng, W. Shi, and X. Shen, "Vehicular communication networks in the automated driving era," IEEE Communications Magazine, vol. 56, no. 9, pp. 26–32, Sept. 2018.
18. Y. Hui, Z. Su, T. H. Luan, and J. Cai, "A game theoretic scheme for optimal access control in heterogeneous vehicular networks," IEEE Transactions on Intelligent Transportation Systems, vol. 20, no. 12, pp. 4590–4603, Dec. 2019.
19. R.I. Meneguette, L.H. Nakamura, A flow control policy based on the class of applications of the vehicular networks, in: Proceedings of the 15th ACM International Symposium on Mobility Management and Wireless Access, MobiWac, 2017, pp. 137–144.
20. H. Ge, Y. Song, C. Wu, J. Ren, G. Tan, Cooperative deep Q-learning with Q-value transfer for multi-intersection signal control, IEEE Access 7 (2019) 40797–40809.
21. X. Liang, X. Du, G. Wang, Z. Han, A deep reinforcement learning network for traffic light cycle control, IEEE Trans. Veh. Technol. 68 (2) (2019) 1243–1253.
22. T. Tan, F. Bao, Y. Deng, A. Jin, Q. Dai, J. Wang, Cooperative deep reinforcement learning for large-scale traffic grid signal control, IEEE Trans. Cybern. (2019) 1–14, Early Access.
23. BMW, Intelligent driving, 2019, [Online]. Available: https://www.bmw.ca/en/topics/experience/connected-drive/bmw-connecteddrive-driver-assistance.html, (Accessed May 2019).
24. Mercedes Benz, Mercedes safety, 2019, [Online]. Available: https://www.mbusa.com/mercedes/benz/safety, (Accessed May 2019).
25. Gaur L, Bhandari M, Razdan T, Mallik S and Zhao Z (2022) Explanation-Driven Deep Learning Model for Prediction of Brain Tumour Status Using MRI Image Data. Front. Genet. 13:822666. doi: https://doi.org/10.3389/fgene.2022.822666

26. Zhao, Y. Chen, L. Lv, L. Deep reinforcement learning with visual attention for vehicle classification, IEEE Trans. Cogn. Dev. Syst. 9 (4) (2017) 356–367.
27. S. Zhang, J. Yang, B. Schiele, Occluded pedestrian detection through guided attention in CNNs, in: Proc. IEEE/CVF CVPR, 2018, pp. 6995–7003.
28. P. Sun, A. Boukerche, Challenges of designing computer vision-based pedestrian detector for supporting autonomous driving, in: Proc. IEEE MASS, 2019, pp. 28–36.
29. G. Brazil, X. Liu, Pedestrian detection with autoregressive network phases, in: Proc. IEEE/CVF CVPR, 2019, pp. 7224–7233.
30. R. Girshick, J. Donahue, T. Darrell, J. Malik, Rich feature hierarchies for accurate object detection and semantic segmentation, in: Proc. IEEE/CVF CVPR, 2014, pp. 580–587.
31. Chao, H.; Cao, Y.; Chen, Y. Autopilots for small unmanned aerial vehicles: A survey. Int. J. Control Autom. Syst. 2010, 8, 36–44.
32. K. C. Santosh and L. Gaur, "Introduction to AI in Public Health," in Artificial Intelligence and Machine Learning in Public Healthcare, Springer, 2021, pp. 1–10.
33. G. Singh, B. Kumar, L. Gaur, and A. Tyagi, "Comparison between Multinomial and Bernoulli Naïve Bayes for Text Classification," in 2019 International Conference on Automation, Computational and Technology Management (ICACTM), 2019, pp. 593–596. doi: https://doi.org/10.1109/ICACTM.2019.8776800.
34. L. Gaur et al., "Capitalising on big data and revolutionary 5G technology: Extracting and visualising ratings and reviews of global chain hotels," Computers & Electrical Engineering, vol. 95, p. 107374, 2021, doi:https://doi.org/10.1016/j.compeleceng.2021.107374.
35. J. Rana, L. Gaur, G. Singh, U. Awan, and M. I. Rasheed, "Reinforcing customer journey through artificial intelligence: a review and research agenda," International Journal of Emerging Markets, vol. ahead-of-print, no. ahead-of-print, Jan. 2021, doi: https://doi.org/10.1108/IJOEM-08-2021-1214.
36. Gaur, L., & Ramakrishnan, R. (2019). Developing internet of things maturity model (IoT-MM) for manufacturing. International Journal of Innovative Technology and Exploring Engineering, 9(1), 2473–2479. doi:https://doi.org/10.35940/ijitee.A4168.119119
37. Gaur, L. (2022). Internet of Things in Healthcare. In: Garg, P.K., Tripathi, N.K., Kappas, M., Gaur, L. (eds) Geospatial Data Science in Healthcare for Society 5.0. Disruptive Technologies and Digital Transformations for Society 5.0. Springer, Singapore. doi:https://doi.org/10.1007/978-981-16-9476-9_6
38. Ramakrishnan, R., Gaur, L. (2016). Application of Internet of Things (IoT) for Smart Process Manufacturing in Indian Packaging Industry. In: Satapathy, S., Mandal, J., Udgata, S., Bhateja, V. (eds) Information Systems Design and Intelligent Applications. Advances in Intelligent Systems and Computing, vol 435. Springer, New Delhi. doi:https://doi.org/10.1007/978-81-322-2757-1_34
39. Gaur, L., Singh, G., & Ramakrishnan, R. (2017). Understanding consumer preferences using IoT smartmirrors. Pertanika Journal of Science and Technology, 25(3), 939–948
40. Singh, G., Gaur, L., & Ramakrishnan, R. (2017). Internet of things – technology adoption model in India. Pertanika Journal of Science and Technology, 25(3), 835–846
41. Singh, G., Gaur, L., & Agarwal, M. (2017). Factors influencing the digital business strategy. Pertanika Journal of Social Sciences and Humanities, 25(3), 1489–1500
42. Ramakrishnan, R., & Gaur, L. (2016). Application of internet of things (iot) for smart process manufacturing in Indian packaging industry doi:https://doi.org/10.1007/978-81-322-2757-1_34
43. Ramakrishnan, R., Gaur, L., & Singh, G. (2016). Feasibility and efficacy of BLE beacon IoT devices in inventory management at the shop floor. International Journal of Electrical and Computer Engineering, 6(5), 2362–2368. doi:https://doi.org/10.11591/ijece.v6i5.10807
44. Ramakrishnan R, Gaur L (2019) Internet of things: approach and applicability in manufacturing, CRC Press 2019.

Chapter 3
Artificial Intelligent Algorithm Based on Energy Efficient Routing for ITS

3.1 Introduction

Varied ad hoc networks, such as vehicular ad hoc networks (VANETs), are considered one of the most promising options for implementing ITS. It is well-known for enhancing road safety. It is true that traffic management and security directly impact the lives of individuals who travel on roadways [1, 2]. An automated monitoring system capable of dealing with traffic congestion and reporting important accident information to healthcare institutions as promptly as feasible is needed. Traffic jams must be avoided, and emergency medical services must get important accident information as fast as possible through the IoT [35–42]. For years, the role of VANETs in ITS has piqued researchers' interest because of their ability to collect, process, and fuse information at low cost while also being fault-tolerant and easy to install at any time desired location [3, 4]. The challenge is creating an effective routing protocol in these networks because of the characteristics, including self-organization, high vehicle mobility, and dynamic topological changes [5]. To improve reliability in VANETs, one of the most important research topics in this field is developing routing protocols that may increase the percentage of packets delivered while minimizing the end-to-end latency. A further problem with VANETs is the security of communication between vehicles themselves. Indeed, because of the distinctive characteristics of VANETs.

In the aviation industry, a UAV, also referred to as a drone, is an aircraft that does not include a human pilot and can fly on its own. Various environmental, military, and commercial applications may be carried out using unmanned aerial vehicles (UAVs), which have flexibility, adaptability, low operating costs, and ease of installation. Sensor's nodes have been installed on roadside units (RSUs) for safety reasons. When an accident occurs, these nodes (fitted with crash sensors) capture the data and transfer it to a sink, from whence it is transmitted for further rescue efforts to the closest traffic information centre (TIC). ITS data communication must be energy efficient since the wireless sensor nodes have a limited battery capacity [6].

UAVs and vehicles may collaborate via heterogeneous communications, which improves the interchange of data between them and benefits various ITS applications, including disaster relief operations [7], remote sensing [8], and other applications. In a metropolis with so many barriers, such as landmarks and buildings, it is possible that the radio signal may be disturbed, resulting in communication between cars failing often. The collaboration of unmanned aerial vehicles (UAVs) with ground vehicles may be investigated considering the properties of VANETs. This routing protocol consists of two routing protocols used to find a route between unmanned aerial vehicles (UAVs) and discover a path between vehicles with the assistance of UAVs. Improved routing algorithms for clustering are achieved via the application of the SHO algorithm. Figure 3.1 demonstrates the EER-SHO operation.

Using UAVs and the proposed EER-SHO protocol, intelligent transportation systems that use VANETs and are equipped to send data packets across the network will better manage their routing processes in the future. When road segments become separated, UAVs help reconnects communication and identify rogue nodes in VANETs [9]. With its investigation and exploitation capabilities, the Spotted Hyena Optimizer (SHO) we developed performs completely. The network performance is predicted to grow if the CH selection is optimised using SHO with more energy-efficient characteristics. The longevity of sensor nodes used to disseminate ITS data is critical and must be emphasized. EER-SHO routing protocol's central purpose is to increase the percentage of packets delivered to healthcare institutions from the roadside and the optimal CH selection.

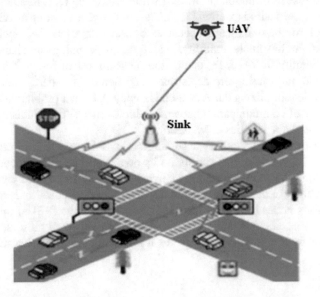

Fig. 3.1 EER-SHO operation demonstration

The significant contribution of this chapter is as follows:

- To enhance the stability period and performance of ad hoc networks, the EER-SHO routing protocol employs a cluster-based approach.
- A sensing system is developed that considers the clustering of nodes distributed along the route. The fitness function of EER-SHO used in the CH selection process considered the factors like residual energy, the distance between node and sink, and intra-cluster distance.
- A mechanism is employed in the EER-SHO protocol to assess the trust value of vehicles, and hostile vehicles are discovered with the assistance of unmanned aerial vehicles (UAVs).
- The suggested routing protocol uses the SHO approach to achieve high packet delivery and network energy while maintaining a long-term connection.

The rest of this chapter is arranged in the following manner: Section 3.2 contains a background study. Section 3.3 outlines the proposed methodology – Sect. 3.4 simulation setting of the suggested routing protocol. Section 3.5 presents simulations results, and Sect. 3.6 concludes the discussion.

3.2   Background Study

VANETs have been the subject of several routing protocol proposals [10, 11]. These protocols try to determine the most efficient path for data transmission to increase throughput and packet delivery ratio while simultaneously minimizing latency [12–15]. Several routing and clustering methods for VANETs are discussed in this literature.

VANET clustering routing protocol QMM-VANET has been suggested by Fatemidokht et al. [16], and it is now under development. This protocol considers the importance of trustworthiness, the needs of quality of service, and the limitations of mobility. To implement the QMM-VANET routing protocol, three stages must be completed: There are three steps in this process: selecting a CH as a vehicle to transmit the packets. QMM-VANET establishes a reliable and trustworthy cluster and improves the connection and stability of communications by increasing their connectivity.

The hybrid technique described by Wang et al. [17] moves the sink optimally by a combination of Ant Colony Optimization (ACO), Differential Evolution (DE), and Particle Swarm Optimization (PSO). On the other hand, the algorithm is vulnerable to the following flaws. The nodes use multi-hop communication because there is no clustering, resulting in the hotspot issue [18]. The number of transmissions is significant due to the lack of clustering. However, since the suggested method has not been tested on the benchmark function, it is prone to performing for extensive portions where it may miss the following optimum location of the sink, which might fail.

Efficient Traffic Light Aware Routing for VANETs is suggested by Oubbati et al. [19]. Traffic density, distance from source to destination, and the current colour of light are all considered when calculating a score for road crossings. However, unlike other protocols, the ETAR protocol does not use the actual distribution of automobiles to determine the degree of connectivity. A data packet may be challenging to carry in certain instances because the network is so sparsely populated.

Secure clustering for ITS is developed by Gaber et al. [20] to create Hierarchical Trust-based Secure Clustering (HiTsec). Aside from the trust values and the number of neighbors that HiTsec considered when selecting its CH, it did not consider distance or other energy-efficient characteristics. It seems that WSNs follow the multi-hop communication model for ITS purposes, leading to hotspots.

Khekare et al. [21] provide ITLs to gather data on traffic density from cars and use it to update the city's congestion data. The cars are then informed of the information that has been gathered. Therefore, vehicles may choose the least congested route possible. Cars equipped with ITLs send out warning signals to nearby vehicles in the case of an accident. In terms of the suggested architecture, the AODV [22] protocol is the best solution since it gives the highest throughput with the least amount of latency.

Bibri et al. [23] included a review and combination of relevant literature to identify and discuss IoT-activated sensor-based big data applications for environmental sustainability and computer models in smart, sustainable cities. An enormous number of active gadgets of all kinds and sizes, with frequent selections being made automatically in particular, facilitates the expansion of urban environments. Smart, sustainable cities are expected to be covered with a layer of electronic technology, interlaced with communication networks and data processing devices.

Carie et al. [24] developed Common Control Channel (CCC) for innovative environments based on the CR-MAC protocol and directional antennas. The Common Control Channel (CCC) is essential for synchronizing nodes, legal access to the channel, and exchanging cognitive messages in control. Comparing the proposed protocol to Omni-directional antennas based on software-defined CR-MAC protocols, the experimental findings reveal that the suggested protocol boosts throughput while decreasing latency and power consumption of the node.

Oubbati et al. [25] presented CRUV protocol that uses unmanned aerial vehicles for VANETs. An acceptable length of time is required to complete the operation. According to this protocol, UAVs and nodes send the messages regularly so that the latter may determine the connection of their nearby segments and improve navigation. All cars stationed at each junction get data packets from the UAVs accessible at each intersection. In addition, when a disconnection occurs, the suggested protocol employs the transport and transmit approach as a mechanism for re-establishing communication.

Using a signature-based method, Golle et al. [26] have developed the concept of comparing the model of legal communications and messages in the VANET environment. Using this method to build a global model of this scope is a disadvantage. In addition, whole communications will be erased from the system. Gurnug et al. [27] introduced a technique that used three metrics to classify received messages into two categories: malicious or lawful announcements. These metrics are the similarity of content, conflict of content, and likeness of the routing path.

Kerrache et al. [28] identified Denial-of-Service attacks via the detection of intrusion approach, which is essential for cybersecurity. When used in DSRC, this approach for discovering DoS attacks takes advantage of the 802.11p access categories to classify the received messages in their early stages, speeding up the detection of intrusions. For the sake of this model, an adversary's dishonest conduct is assumed to be consistent throughout time. Consequently, it does not initiate a viable remedy when nodes engage in intelligent, malevolent behavior.

To identify hostile vehicles in VANETs, Kerrache and colleagues [29] developed a trust model based on the assistance of unmanned aerial vehicles (UAVs). It comprises a discovery threshold approach that allows for the detection of intelligent harmful conduct, such as the modification or faking of identification techniques, among other things. A clustering strategy based on unmanned aerial vehicles (UAVs) is also included in the suggested approach, which reduces the number of messages exchanged while also storing the energy of UAVs. It is possible to evaluate the level of trust between vehicles using this technique. After that, unmanned aerial vehicles (UAVs) classify road segments into static clusters and begin a phase of data collection. The detection and packet delivery ratios may be enhanced by using the suggested approach.

Using a Fuzzy Classification Trust Model (FCTM), Singh and Verma [30] have developed a framework for FANETs. The node's behavior and cooperation in the operation of the network are considered when applying this technique. Additionally, they use social characteristics and QoS to increase the trustworthiness of each unmanned aerial vehicle. The suggested strategy is called Information-Centric Networks (ICN).

Yu et al. [31] developed an ACO-based polymorphism-aware routing technique for FANETs, which is based on polymorphism-aware routing (PAR). DSR (Dynamic Source Routing) techniques are integrated into the proposed algorithm, allowing it to perform better than each of them independently. Congestion, the distance between places on the route, and route stability are all considered to determine a route's pheromone value. It is also possible to provide a new mechanism for pheromone evaporation. When compared to existing methods, the APAR algorithm has the potential to enhance packet delivery ratio, routing overhead, and end-to-end latency in the combat environment, among other things [34].

Bensalem and Boubiche [32] have presented a revolutionary ElectriBio-inspired Energy-Efficient Self-organization model for Unmanned Aerial Ad-hoc Networks, and it is described in detail below (EBEESU). The primary purpose of EBEESU is to save energy, which is critical to the longevity of unmanned aerial vehicles (UAVs). As a result, EBEESU is a mix of biologically inspired and electrically inspired models and an algorithm for cluster-based communication with two levels of data aggregation.

3.3 Proposed Work: EER-SHO

We investigate on VANET routing protocol that uses unmanned aerial vehicles in conjunction with a group of automobiles to enhance routing. When the network on the ground is sparsely linked, this protocol uses unmanned aerial vehicles for several different activities, including data delivery to a specific location, computing road segment connectivity, and maintaining the network. Because connection failure is a typical occurrence in VANETs, collaboration between vehicles and unmanned aerial vehicles (UAVs) may aid in reducing the delay in delivery and packet losses. In general, two types of data delivery methods may be used in parallel: delivery of data packets via CHs as the vehicles and unmanned aerial vehicles and data delivery through communication between unmanned aerial vehicles.

Dhiman et al. [33] proposed SHO by taking inspiration from spotted hyenas' encircling, attacking, hunting, and searching mechanism. The encircling behaviour is mathematically defined as follows:

$$\overrightarrow{E_h} = \# \ \vec{C} \cdot \overrightarrow{Y_p} \ (t) - \vec{Y}(t) \# \tag{3.1}$$

$$\vec{Y}(t+1) = \overrightarrow{Y_p}(t) - \vec{F} \cdot \overrightarrow{E_h} \tag{3.2}$$

where, $\overrightarrow{E_h}$ indicates the distance between spotted hyena and prey, and t signifies the current iteration. The position vectors of prey and spotted hyena are represented by $\overrightarrow{Y_p}$, and $\vec{Y}$, respectively. Furthermore, the symbols · and || refer to the multiplication vector and absolute value, respectively. The coefficient vectors $\vec{C}$ and $\vec{F}$ are computed mathematically as:

$$\vec{C} = 2 \cdot \overrightarrow{r_1} \tag{3.3}$$

$$\vec{F} = 2\vec{h} \cdot \overrightarrow{r_2} - \vec{h} \tag{3.4}$$

$$\vec{h} = 5 - \left(Itr + \frac{5}{\max_{Itr}} \right) \tag{3.5}$$

where $Itr = 0, 1, 2, Max_{Itr}$.

Where, $\overrightarrow{r_1}$ and $\overrightarrow{r_2}$ are random vectors with values lying in the [0, 1] domain. On manipulating $\vec{F}$ and $\vec{C}$, hyenas can reach several different locations. Moreover, vector $\vec{h}$ is decremented from five to zero over the passage of iterations. Furthermore, spotted hyenas usually live and chase in clusters of reliable companions and own the ability to locate prey's position. The below-mentioned equations are employed to determine the potential hunting area:

$$\overrightarrow{E_h} = \# \ \vec{C} \cdot \overrightarrow{Y_h} - \overrightarrow{Y_r} \ \# \tag{3.6}$$

$$\overrightarrow{Y_r} = \overrightarrow{Y_h} - \vec{F} \cdot \overrightarrow{E_h} \qquad (3.7)$$

$$\overrightarrow{B_h} = \overrightarrow{Y_r} + \overrightarrow{Y_{r+1}} + \ldots + \overrightarrow{Y_{r+i}} \qquad (3.8)$$

Where, the placement of the best search agent and other spotted hyenas is indicated by $\overrightarrow{Y_h}$ and $\overrightarrow{Y_r}$, respectively. I signifies the count of spotted hyenas and is estimated as:

$$I = count_{ns}\left(\overrightarrow{Y_h}, \overrightarrow{Y_{h+1}}, \ldots, \left(\overrightarrow{Y_h} + \vec{U}\right)\right) \qquad (3.9)$$

$$\vec{Y}(t+1) = \frac{\overrightarrow{B_h}}{I} \qquad (3.10)$$

where, $\vec{U}$, $count_{ns}$, $\overrightarrow{B_h}$, and $\vec{Y}(t+1)$ indicate a random vector having values in the range [0.5, 1], the count of candidate solutions, and the bundle of optimal solutions and best optimal solution, respectively.

After the termination requirement is fulfilled, the best solution available at the time is chosen.

The proposed EER-SHO method determines the fitness population which node is suited for becoming CH. The fitness function determines the fitness of each person, which is the node in this case and provides the best strategy for maintaining the nodes' energy. The importance of fitness factors, in this case, should not be overlooked. It is critical to provide those crucial fitness characteristics that will ultimately determine whether or not CH is chosen. The fitness parameters: We use three essential fitness metrics to help us choose the right CH.

Residual Energy: The remaining energy value of the node is the essential metric. Compared to the other nodes, the CH uses energy at a greater rate. As a result, the node with the highest energy value must be chosen. All nodes start with the initial energy, but it depletes at varying rates based on their distance from the sink. As a result, the remaining energy is considered when choosing a CH. The remaining energy sensor node is determined by Eq. (3.13), which is added together to get the residual energy of all nodes.

$$f_1 = \frac{1}{N} \sum_{N}^{i=1} E_{(i)} \qquad (3.11)$$

$$f_2 = \frac{1}{cl} \sum_{cl}^{i=1} E_{(n)} \qquad (3.12)$$

$$Obj_1 = \frac{f_1}{f_2} \qquad (3.13)$$

Distance Between Sink and Node: The energy is utilized to determine the distance between member nodes and the sink. The sum of energy consumed by the sink remains strictly proportional to the distance from the sink to the node. Therefore, the networking technique for CH selection considers the factors under the median gap between the member nodes, and the base station can be optimized accordingly. The target purpose rather than CH selection decisions should be framed, Obj_2 is presented in conjunction with the Eq. (3.14) with the distance variable.

$$Obj_2 = \sum_{N}^{i=1} \left(\frac{D_{(N(i)-S)}}{D_{AVG(N(i)-S)}} \right) \tag{3.14}$$

where,

$$D_{AVG(N(i)-S)} = \left(\frac{\sum_{i=1,}^{N} D_{(N(i)-S)}}{N} \right) \tag{3.15}$$

The above Eq. (3.14) evaluates the sum of the distance costs obtained for any ith node by the third fitness parameter (Obj_2), where i ranges from 1 to M (Number of nodes) and in the Eq. (3.15), $D_{N(i)-S}$ denotes the Euclidean distance from the sink to the ith node, $D_{AVG(N(i)-S)}$ signifies an average distance at the middle of the ith node and sink.

Intra-Cluster Distance: Those nodes with more energy and closer to the central station have a better chance of becoming cluster heads. Also, when it comes to inter-cluster and intra-cluster distances, the cluster heads are more evenly distributed across the whole network, which is better for everyone. The new method tries to keep the average distances between cluster members and cluster heads as small as possible. There should also be much space between the cluster heads.

$$Obj_3 = \sum_{N}^{i=1} \left(\frac{D_{(N(i)-S)}}{D_{AVG(N(i)-S)}} \right) * \frac{1}{0.1M} \tag{3.16}$$

CHs and member nodes are found once each particle is selected for the standard clustering process. Each member node is assigned to the nearest cluster head in the following step, forming clusters. The suggested purpose function is used to calculate the error degree of each of the k members of the population.

Fitness function: We considered above that the network's integration of various fitness parameters is wholly expressed in each fitness function as follows in Eq. (3.17).

$$F = \alpha * Obj_1 + \beta * Obj_2 + \gamma * Obj_3 \tag{3.17}$$

Minimizing fitness F in Eq. (3.17) is necessary to improve network performance. Includes various fitness parameters, which are computed in Eqs. (3.11), (3.12), (3.13), (3.14), (3.15) and (3.16). In this Eq. (3.17), the weight coefficients give different weightage to the different parameters used in the integration of fitness function. It depends on the user to tune these parameters according to the sensor network's application.

Algorithm 3.1: EER-SHO Method

```
Input: Network formation parameters
Output: O = CH_FSL,
 1: CH_FSL = 0.
 2: for Roundcount = 1 to Roundmaxm do
 3:       Al_n = N
 4:       Dd_n= 0
 5:       for j= 1 to T_tN do
 6:             if Er(p) == 0 then
 7:                   Dd_n=Dd_n+1
 8:                   if Dd_n==n then
 9:                         Al_dead=RCurr
10:                   end if
11:                   Al_n = Al_n - Dd_n
12:             end if
13:       end for
14:       for k= 1 to Tnd do
15:       if Er(k) > 0 then
16:             Computing fitness function through SHO
17:             if F(k) is having maximum value then
18:                   Select k^th node as CH
19:                   CH_FSL = CH_FSL + 1
20:                   CH_FSL ← CDMA slot
21:                   CH_FSL ← k^th (data tx)
22:             else
23:             k^th node is non-CH (cluster member)
24:             k^th node ← assignment of TDMA slot
25:             end if
26:       Update Erm(k)
27:       else
28:             break
29:       end if
30:       end for
31:       if Dd_n==T_tN then
32:             break
33:       end if
34: end for
35: return O
```

Algorithm 3.1 outlines the many processes considered while running EER-SHO. The algorithm's starting parameters are the total number of nodes n and the maximum number of iterations the suggested approach is applicable. This algorithm's result is a choice of CH based on energy-efficient routing utilizing SHO [43].

3.4   Simulation Setting

The performance of the proposed EER-SHO was evaluated on the MATLAB platform, which is chosen for this purpose. During the simulation, we dispersed 200 energy-enable sensor nodes throughout a two-dimensional region (200 m × 200 m) to capture the most energy possible in VANET. We also set up a sink node (100 m, 100 m). Each sensor node is fitted with a rechargeable battery with a maximum capacity of 200 Joules. The simulation of the suggested technique and various outcome parameters is shown in this part, which is done in MATLAB.

On the other hand, we have compared the EER-SHO against other protocols such as AoDV, HiTsec, and DOCAT. Many types of experiments use the EER-SHO algorithm, including a run-time instance of the method, network lifespan, network stability period, cumulative data packet transmitted, and energy-consuming rate. Table 3.1 contains a complete list of all the parameters used in the simulations.

Table 3.1 Simulation parameters

Parameters	Values
Coverage area size	200 m × 200 m
Sensor nodes	200
Sink node	1
Initial energy *(Eo)*	1 J
Essential transceiver energy (E_{el})	50 nJ/bit
UAV	1
Size of the data packets	2000bits
Utilization of energy in data aggregation *(E_{da})*	5 nJ/bit/signal
α, β, and γ	0.55, 0.35, and 0.2
Simulation runs	20
Frequency band	4.5 GHz

3.5   Simulation Results

Standard performance metrics are used to validate the performance of the proposed EER-SHO. Four performance metrics, such as the network's remaining energy, network longevity, throughput, and stability period considered for the selection of CH. These metrics are defined and explained for the EER-SHO against the protocols: AoDV, HiTsec, and DOCAT.

Network Stability Period: The duration between the start of a network and the death of its first node is known as the network's stability period. The stability period is critical for ensuring overall network operations with effective data transmission for a longer time. Figure 3.2 shows that the suggested method has the most extended stability period of other existing protocols. In EER-SHO, the network completed 11,615 rounds, and other protocols AoDV, HiTsec, and DOCAT were completed in 3687, 5315, and 7837 rounds. An improvement such as stability period and HND is the unification of three fitness factors that make sure the energy conservation even in data transmission. Therefore, the distance between the nodes and sink to nodes is efficiently decreased.

Network's Remaining Energy: It is shown in Fig. 3.3 that the overall network energy is 200 J, which rapidly decreases as the rounds. This metric helps determine the rate at which different protocols use energy. As shown in Fig. 3.3, the network begins to use energy as soon as the EER-SHO operation begins, and it covers more

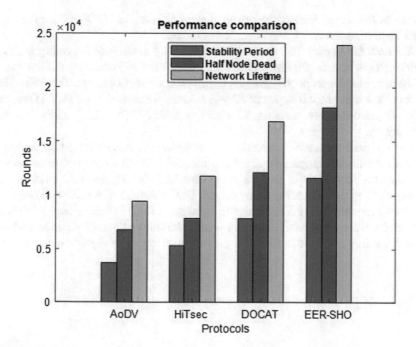

Fig. 3.2 Performance analysis of EER-SHO and other protocols

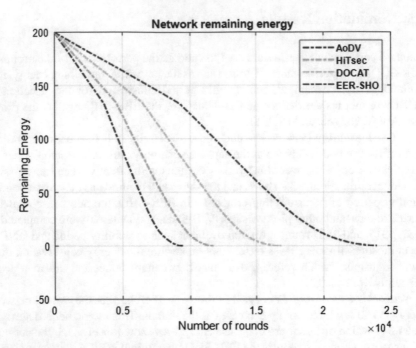

Fig. 3.3 Remaining energy analysis of EER-SHO and other protocols

rounds as the nodes' energy decreases. AoDV, HiTsec, and DOCAT, on the other hand, require more energy and cover fewer circles.

Network Longevity: The network lifespan is the number of surviving nodes at various cluster rounds. Figure 3.4 illustrates the results of comparing the number of living nodes of the proposed protocol to different current systems. The EER-SHO protocol has a network lifespan of 23,943 rounds, whereas the AoDV, HiTsec, and DOCAT protocols have network lifetimes of 9426, 11,765, and 16,844 rounds, respectively.

Throughput: Figure 3.5, the rate of data packets is successfully transmitted to the base station. In comparison to other protocols, EER-SHO outperforms them by sending a higher number of packets to the sink. EER-SHO sends 1,306,818 packets to the sink, whereas AoDV, HiTsec, and DOCAT send 474,687, 606,668, and 880,114 packets to the sink. The higher throughput value is that critical criteria for CH selection are considered, which improves the value of stability period and network lifespan. As a result, EER-SHO performance improves significantly.

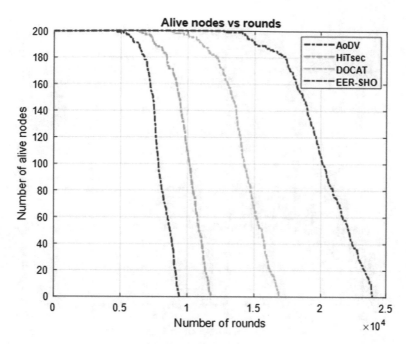

Fig. 3.4 Alive nodes analysis of EER-SHO and other protocols

3.6 Conclusion

An energy-efficient routing utilization SHO enhances the energy consumption of VANET. Due to its rapid convergence and increased variety, SHO is used to improve the CH selection in this chapter proposed sensing system, dubbed EER-SHO. SHO employs a fitness function for each of the CH selection fitness parameters. When an accident occurs on the road, the sink at EER-SHO collects data that might be used in medical settings such as emergency response centers. A series of MATLAB simulations show that EER-SHO is computationally efficient. This method improves the network's total residual energy and helps maintain the network's stability and throughput for the longest time possible compared to AoDV, HiTsec, and DOCAT methods. Another significant concern that needs to be optimized is the energy-enabled nodes should be used. Furthermore, it will observe that the network performance improves when the sink moves in the energy harvesting VANET.

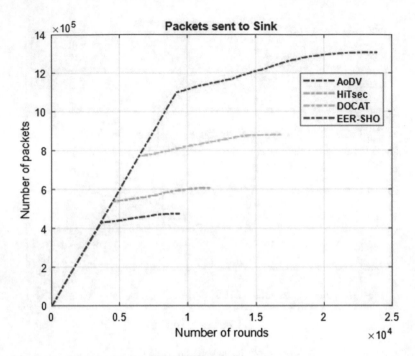

Fig. 3.5 Packet delivery analysis of EER-SHO and other protocols

References

1. G. Dimitrakopoulos and P. Demestichas, "Intelligent transportation systems," IEEE Veh. Technol. Mag., vol. 5, no. 1, pp. 77–84, Mar. 2010
2. Sultan, M. M. Al-Doori, A. H. Al-Bayatti, and H. Zedan, "A comprehensive survey on vehicular ad hoc network," J. Netw. Comput. Appl., vol. 37, pp. 380–392, Jan. 2014
3. D. Tacconi, D. Miorandi, I. Carreras, F. Chiti, and R. Fantacci, "Using wireless sensor networks to support intelligent transportation systems," Ad Hoc Netw., vol. 8, no. 5, pp. 462–473, Jul. 2010.
4. M. J. Piran, A. Ali, and D. Y. Suh, "Fuzzy-based sensor fusion for cognitive radio-based vehicular ad hoc and sensor networks," Math. Problems Eng., vol. 2015, Mar. 2015, Art. no. 439272.
5. B. T. Sharef, R. A. Alsaqour, and M. Ismail, "Vehicular communication ad hoc routing protocols: A survey," J. Netw. Comput. Appl., vol. 40, pp. 363–396, Apr. 2014.
6. I. Bekmezci, O. K. Sahingoz, and S. Temel, "Flying ad-hoc networks (FANETs): A survey," Ad Hoc Netw., vol. 11, no. 3, pp. 1254–1270, May 2013.
7. E. Yanmaz, M. Quaritsch, S. Yahyanejad, B. Rinner, H. Hellwagner, and C. Bettstetter, "Communication and coordination for drone networks," in Proc. Ad Hoc Netw. Ottawa, ON, Canada: Springer, 2017, pp. 79–91.
8. R. Ramakrishnan, L. Gaur, Smart electricity distribution in residential areas: Internet of things (IoT) based advanced metering infrastructure and cloud analytics, Internet of Things and Applications (IOTA), International Conference on, IEEE (2016), pp. 46–51
9. S. A. Hadiwardoyo, E. Hernández-Orallo, C. T. Calafate, J. C. Cano, and P. Manzoni, "Experimental characterization of UAV-to-car communications," Comput. Netw., vol. 136, pp. 105–118, May 2018.

10. A. Daeinabi, A. G. P. Rahbar, and A. Khademzadeh, "VWCA: An efficient clustering algorithm in vehicular ad hoc networks," J. Netw. Comput. Appl., vol. 34, no. 1, pp. 207–222, Jan. 2011.
11. K. Liu, J. Zhang, and T. Zhang, "The clustering algorithm of UAV networking in near-space," in Proc. 8th Int. Symp. Antennas, Propag. EM Theory, 2008, pp. 1550–1553.
12. Sahoo BM, Amgoth T, Pandey HM. Particle swarm optimization based energy efficient clustering and sink mobility in heterogeneous wireless sensor network. Ad Hoc Networks. 2020 Sep 1;106:102237.
13. Sahoo, Biswa Mohan, and Tarachand Amgoth. "An Improved Bat Algorithm for Unequal Clustering in Heterogeneous Wireless Sensor Networks." SN Computer Science 2, no. 4 (2021): 1–10.
14. Sahoo BM, Pandey HM, Amgoth T. GAPSO-H: A hybrid approach towards optimizing the cluster-based routing in wireless sensor network. Swarm and Evolutionary Computation. 2021 Feb 1;60:100772.
15. Sahoo BM, Pandey HM, Amgoth T. A whale optimization (WOA): Meta-heuristic based energy improvement clustering in wireless sensor networks. In2021 11th International Conference on Cloud Computing, Data Science & Engineering (Confluence) 2021 Jan 28 (pp. 649–654). IEEE.
16. H. Fatemidokht and M. K. Rafsanjani, "QMM-VANET: An efficient clustering algorithm based on QoS and monitoring of malicious vehicles in vehicular ad hoc networks," J. Syst. Softw., vol. 165, Jul. 2020, Art. no. 110561.
17. H. Wang, K. Li, and W. Pedrycz, "An elite hybrid Metaheuristic optimization algorithm for maximizing wireless sensor networks lifetime with a sink node," IEEE Sensors J., vol. 20, no. 10, pp. 5634–5649, May 2020.
18. Sahoo BM, Amgoth T, Pandey HM. Enhancing the Network Performance of Wireless Sensor Networks on Meta-heuristic Approach: Grey Wolf Optimization. In Applications of Artificial Intelligence and Machine Learning 2021 (pp. 469–482). Springer, Singapore.
19. O. S. Oubbati, A. Lakas, N. Lagraa, and M. B. Yagoubi, "ETAR: Efficient traffic light aware routing protocol for vehicular networks," in Proc. Int. Wireless Commun. Mobile Comput. Conf. (IWCMC), Aug. 2015, pp. 297–301.
20. T. Gaber, S. Abdelwahab, M. Elhoseny, and A. E. Hassanien, "Trust based secure clustering in WSN-based intelligent transportation systems," Comput. Netw., vol. 146, pp. 151–158, Dec. 2018.
21. G. S. Khekare and A. V. Sakhare, "A smart city framework for intelligent traffic system using VANET," in Proc. Int. Mutli-Conf. Autom., Comput., Commun., Control Compressed Sens. (iMac4s), Kottayam, India, Mar. 2013, pp. 302–305.
22. C. Perkins, E. Belding-Royer, and S. Das, "Ad hoc on-demand distance vector (AODV) routing," RFC Editor, USA, Tech. Rep. RFC3561, 2003.
23. S. E. Bibri, "The IoT for smart sustainable cities of the future: An analytical framework for sensor-based big data applications for environmental sustainability," Sustain. Cities Soc., vol. 38, pp. 230–253, Apr. 2018.
24. A. Carie et al., "An Internet of software defined cognitive radio ad-hoc networks based on directional antenna for smart environments," Sustain. Cities Soc., vol. 39, pp. 527–536, May 2018.
25. O. S. Oubbati, A. Lakas, N. Lagraa, and M. B. Yagoubi, "CRUV: Connectivity-based traffic density aware routing using UAVs for VANets," in Proc. Int. Conf. Connected Vehicles Expo (ICCVE), Oct. 2015, pp. 68–73.
26. P. Golle, D. Greene, and J. Staddon, "Detecting and correcting malicious data in VANETs," in Proc. 1st ACM Workshop Veh. Ad Hoc Netw. (VANET), Philadelphia, PA, USA, 2004, pp. 29–37.
27. S. Gurung, D. Lin, A. Squicciarini, and E. Bertino, "Information oriented trustworthiness evaluation in vehicular ad-hoc networks," in Proc. Int. Conf. Netw. Syst. Secur. Sapporo, Japan: Springer, 2013, pp. 94–108.

28. C. A. Kerrache, N. Lagraa, C. T. Calafate, and A. Lakas, "TFDD: A trust-based framework for reliable data delivery and DoS defense in VANETs," Veh. Commun., vol. 9, pp. 254–267, Jul. 2017.
29. C. A. Kerrache, A. Lakas, N. Lagraa, and E. Barka, "UAV-assisted technique for the detection of malicious and selfish nodes in VANETs," Veh. Commun., vol. 11, pp. 1–11, Jan. 2018.
30. K. Singh and A. K. Verma, "FCTM: A novel fuzzy classification trust model for enhancing reliability in flying ad hoc networks (FANETs)." Ad Hoc Sensor Wireless Netw., vol. 40, nos. 1–2, pp. 23–47, 2018.
31. Y. Yu, L. Ru, W. Chi, Y. Liu, Q. Yu, and K. Fang, "Ant colony optimization based polymorphism-aware routing algorithm for ad hoc UAV network," Multimedia Tools Appl., vol. 75, no. 22, pp. 14451–14476, Nov. 2016.
32. A. Bensalem and D. E. Boubiche, "EBEESU: Electric Bio-inspired energy-efficient self-organization model for unmanned aerial ad-hoc network," Ad Hoc Netw., vol. 107, Oct. 2020, Art. no. 102236.
33. Dhiman G, Kumar V. Spotted hyena optimizer: a novel bio-inspired based metaheuristic technique for engineering applications. Advances in Engineering Software 2017; 114: 48–70.
34. Nayak SR, Sahoo BM, Malarvel M, Mishra J, editors. Smart Sensor Networks Using AI for Industry 4.0: Applications and New Opportunities. CRC Press; 2021 Oct 10.
35. Gaur, L. (2022). Internet of Things in Healthcare. In: Garg, P.K., Tripathi, N.K., Kappas, M., Gaur, L. (eds) Geospatial Data Science in Healthcare for Society 5.0. Disruptive Technologies and Digital Transformations for Society 5.0. Springer, Singapore. doi:https://doi.org/10.1007/978-981-16-9476-9_6
36. Ramakrishnan, R., Gaur, L. (2016). Application of Internet of Things (IoT) for Smart Process Manufacturing in Indian Packaging Industry. In: Satapathy, S., Mandal, J., Udgata, S., Bhateja, V. (eds) Information Systems Design and Intelligent Applications. Advances in Intelligent Systems and Computing, vol 435. Springer, New Delhi. doi:https://doi.org/10.1007/978-81-322-2757-1_34
37. Gaur, L. & Ramakrishnan, R. Developing internet of things maturity model (IoT-MM) for manufacturing. Int. J. Innovative Technol. Exploring Eng., 2019. doi: https://doi.org/10.35940/ijitee.A4168.119119.
38. Gaur, L., Singh, G., & Ramakrishnan, R. (2017). Understanding consumer preferences using IoT smartmirrors. Pertanika Journal of Science and Technology, 25(3), 939–948
39. Singh, G., Gaur, L., & Ramakrishnan, R. (2017). Internet of things – technology adoption model in India. Pertanika Journal of Science and Technology, 25(3), 835–846
40. Singh, G., Gaur, L., & Agarwal, M. (2017). Factors influencing the digital business strategy. Pertanika Journal of Social Sciences and Humanities, 25(3), 1489–1500
41. Ramakrishnan, R., & Gaur, L. (2016). Application of internet of things (iot) for smart process manufacturing in Indian packaging industry https://doi.org/10.1007/978-81-322-2757-1_34
42. Ramakrishnan, R., Gaur, L., & Singh, G. (2016). Feasibility and efficacy of BLE beacon IoT devices in inventory management at the shop floor. International Journal of Electrical and Computer Engineering, 6(5), 2362–2368. doi:https://doi.org/10.11591/ijece.v6i5.10807
43. Ramakrishnan R, Gaur L (2019) Internet of things: approach and applicability in manufacturing, CRC Press 2019 Jun 4

Chapter 4
Intelligent Transportation System: Modern Business Models

4.1 Introduction

From 2021 to 2028, the global intelligent transportation system market is estimated to increase at a CAGR of 7.0%, from USD 25,378.2 million in 2020 to USD 25,378.2 million in [1] 2028. The market is predicted to increase due to rising demand for traffic control solutions, intelligent vehicles connected with cutting-edge [2] telecommunications technology, and improved safety and surveillance provided by sophisticated cameras and License Plate Recognition (LPRs) [3–5]. Other reasons expected to contribute to market growth include an increase in the number of vehicles on the road, ageing infrastructure, and a looming shortage of traffic data management. The pressing necessity to effectively regulate traffic flow across metropolitan corridors and highways [6] has prompted the development of a new traffic management system. As a result, transportation authorities anticipate using cutting-edge data analytics tools to address [7] the issue of rising traffic.

Modern business models for Intelligent Transportation Systems (ITS).

Mobility, safety, and environmental applications [8] are the three primary categories of ITS applications.

ITS are a collection of cutting-edge information and communication technologies used in transportation and traffic management systems to improve transportation networks' safety, efficiency, and [9] sustainability, reduce traffic congestion, and improve driver experiences [10].

© The Author(s), under exclusive license to Springer Nature Switzerland AG 2022   67
L. Gaur, B. M. Sahoo, *Explainable Artificial Intelligence for Intelligent
Transportation Systems*, https://doi.org/10.1007/978-3-031-09644-0_4

4.2   ITS and Economic Growth

Socioeconomic factors influence the development of transportation systems. Recent years have seen a better balance between development policies and strategies [11–15], including human capital issues. However, significant physical capital may be compared to human capital; both must interact to facilitate development, just as infrastructures cannot function properly without proper operation and maintenance. On the other hand, economic activities are impossible without sound infrastructure. Many transport activities are highly transactional and service-oriented, underscoring the complex relationship between physical and human capital needs. An effective logistics system, for example, relies on sophisticated infrastructure and management skills [16]. In addition to being an essential part of the economy, a transportation sector is a standard tool for development due to its intensive use of infrastructures. The economic opportunity to move people and freight [17–22] and to use information and communication technologies has become even more critical because of the global economy. Economic development is related to the amount and quality of transport infrastructure. A high level of development is often associated with dense transport infrastructure [23–28] and well-connected networks. Having efficient transportation systems results in economic and social benefits, which lead to greater accessibility to markets, a more robust economy, and more investment opportunities. In the absence of reliable [29–32] transportation systems, cost-related consequences can include lost opportunities and reduced quality of life.

The expense of inefficient transportation increases in many economic sectors, whereas the cost of efficient transportation reduces these costs. Transportation impacts can also have unforeseen or unintended consequences since they are not always intended. Congestion is often unintentional when free or low-cost transport infrastructure is provided to users [33–36]. A growing economy indicated by congestion can also demonstrate a limited capacity and an insufficient infrastructure to handle the rising mobility demands. We mustn't ignore the social and environmental costs of transportation.

In ITS, various technologies provide optimal technical equipment for vehicles to efficiently utilise transport infrastructure. Road users and operators can access real-time information that improves safety and [37] mobility while reducing the environmental impact of transportation. Globally, several authorities and operators have adopted modern technologies, like ITS, to improve the management of transportation networks.

4.2.1   COVID-19 and ITS

Due to supply chain bottlenecks and component shortages, the COVID-19 epidemic has significantly impacted new ITS sales and installations. Nonetheless, existing ITS solutions installed around the world are being actively used. The introduction

```
┌─────────────────────────────┐   ┌───────────────────────────────────┐
│                             │   │           Application             │
│                             │   │        Traffic Management         │
│  Intelligent Transport System│   │     Road Safety & Surveillance    │
│      Market Segmentation    │   │        Fright Management          │
│                             │   │         Public Transport          │
│                             │   │       Environment Protection      │
└─────────────────────────────┘   │         Automated Vehicle         │
                                   │        Parking Management         │
                                   └───────────────────────────────────┘

┌──────────────────────────────────────┐   ┌───────────────────────────┐
│              System type             │   │          Region           │
│  Advanced Traveller Information System (ATID)│ │       North America       │
│ Advanced Transportation Management System (ATMS)│ │         Europe          │
│  Advanced Transportation Pricing System (ATPS)│ │       Asia Pacific        │
│   Advanced Public Transportation System (APTS)│ │   Middle East & Africa    │
│      Co-operative Vehicle System (CVS)│   │       Latin America        │
│ Automatic Number Plate Recognition System (ANPR)│ └───────────────────────────┘
└──────────────────────────────────────┘
```

Fig. 4.1 Represents the segmentation of the ITS market

of intelligent technologies has proved the role of technology in molding the future, even though private and [38] public initiatives have been adequate to various degrees. Lockdowns, border closures, and other restrictions implemented by different countries worldwide to stop coronavirus spread have taken a heavy toll on supply chains. Manufacturing and assembling facilities have also been temporarily shut down. On the other hand, ongoing transportation projects are postponed due to a lack of staff (Fig. 4.1).

At this point, the ITS ecosystem, which is heavily reliant on electronics and telecommunications, is expected to decelerate due to supply chain interruptions. A delay in the deployment of ITS appears to be unavoidable in the short term.

4.3   Impact of ITS Advances on the Industry

A typical ITS collects real-time data using a combination of sensors, video surveillance devices, navigation systems, and vehicle probes, which are then evaluated and disseminated to users via the internet, mobile telephony, or dynamic signboards. As a result, ITS applies to rail, road, water, and air transportation. Over the projection period, advances in sensing and telecommunications technologies are expected to increase the deployment of ITS. The growing inclination for adopting an ITS, on the other hand, is emerging as one of the primary elements driving the digitisation of many aspects of the transportation infrastructure ecosystem.

The deployment of ITS to reduce traffic accidents and improve road safety is a crucial driver driving the intelligent transportation system market forward. Over the next few years, the need for Vehicle-to-Infrastructure (V2I) and Vehicle-to-Vehicle (V2V) communication to improve road safety is likely to grow. Ongoing advances in the transportation industry have spurred the demand for an efficient transportation system that might improve road networks. Simultaneously, technological advancements such as blind-spot detection and computerised toll collection have reshaped expectations and prospects for sustainable transportation and traffic management.

The growing number of vehicles on the road and the resulting frequent traffic jams necessitate the implementation of advanced traffic management technologies. These solutions can shorten travel times while also assuring efficient traffic management. Authorities and public safety organisations can also use advanced traffic management systems to respond quickly and effectively to accidents and emergencies.

As part of their attempts to grow their respective market shares, market players aggressively invest in research and development operations, upgrade internal processes, actively engage in new product creation, and improve their existing products. To develop technologically innovative products and acquire a competitive edge globally, they also focus on strategic partnerships and mergers and acquisitions. Denso Corporation, for example, teamed with Aeva, a producer of LiDAR and perception systems for autonomous vehicles, to develop sensing and perception systems in January 2021.

The two businesses will work together to develop Frequency Modulated Continuous Wave (FMCW) LiDAR, which could aid in assessing vehicle velocity and reflectivity. Market participants are investing heavily in research and development to fuel organic growth. For example, Hitachi, Ltd. introduced a heavy-duty automated parking brake (APB HD) for vans, SUVs, pickup trucks, and light commercial vehicles in October 2020. The new approach improves APB safety and system administration while allowing for more excellent technology applications.

The region's most advanced market for ITS in North America has multiple; ITS systems in place to handle the growing traffic and public transportation system. Commercial drone sales are soaring throughout the region, with countries like the United States enacting strict regulations for drone pilots and registration. Consequently, the demand for ITS systems in the aviation sector is predicted to rise. North America has a considerable demand for dedicated short-range communication devices that reduce traffic congestion in passenger and commercial vehicles, boosting the demand for ITS systems in the region.

4.4   ITS Business Models: Ticketing as a Service

The model describes how a company runs and generates value. It forges a link between an organisation's vision, structures, and operations. In the 1990s, IBM invented the term "e-Company model," which it defined as "the transformation of core business activities through internet technology." It incorporates online customer service, and corporate collaboration, making it broader than e-Commerce. Despite the apparent emphasis on commercial sustainability, and sustainable business models in the ITS industry must consider the sector's diverse nature and distinctive characteristics. There is no universally accepted definition of a business or an e-Business model. Several authors have investigated e-Business models and created taxonomies based on standard criteria. Although it is difficult to determine what should and should not be included in an eBusiness model, and there is no such thing as a universal business model, Osterwalder and Pigneur developed one of the most comprehensive e-Business models frameworks ever. They propose an e-Business model based on different pillars, including internal and external stakeholders, which is crucial to any business model's long-term viability. These challenges must be solved enough for any ITS project to be sustainable. ITS is focused on long-term e-business concepts.

4.4.1. Services and products: Offerings, services or products, are the foundation for creating revenue, and people from all walks of life feel their consequences. The ITS value proposition is offered. Creating value begins with ITS suppliers providing intelligent vehicles and infrastructure. The value created is passed on to consumers regarding safety, decreased travel time and reduced congestion. This layer includes the direct benefits that consumers are willing to pay. As a result, more organisations will be able to enter the market as mediators and providers of services. The value is subsequently transferred to a higher layer, the "Economy & Society." There are various external benefits (also known as positive economic externalities) associated with reduced investment in new roads. These benefits include improved traffic management, less congestion, lower expenses, and a decrease in the requirement for accident response and treatment. It also helps to minimise pollution, increase social inclusion for vulnerable transit users like elderly and disabled people and improve the general quality of life for all people involved.

4.4.2. Intelligent infrastructure: Governments, financial institutions, transportation organisations, automobile manufacturers, information and communication technology enterprises, the energy sector, road users, and others could all be part of an ITS network. The next part digs into the ITS research from the stakeholders' perspective. It all boils down to how to best utilise intelligent infrastructure to generate value for the organisation. It also defines the primary partners, their roles, their goals and the various types of interactions and partnerships that may be conceivable in the future. The internet allows businesses to establish e-Business models requiring minor physical infrastructure and manage their relationships with business partners. Firms that manipulate digital road infrastructure services may be able to benefit from these prospects. However, issues relating to vehicle compatibility must be given special consideration. According to the industry, intelligent vehicle infrastructure that is incompatible with ITS technology adds little value and wastes money. Because of this, intelligent cars should be viewed as part of the infrastructure when adopting a specific eBusiness model rather than as a standalone component. The third component, relationship capital, is also essential because it builds a customer network and manages client relationships.

4.4.3. Customer relationship: It is concerned with customer relationships and the impact of ITS networks on road users. It also addresses how ITS providers can earn users' trust in their products and services, how trust can be measured, how users' information is collected and used, and how service providers gather and manipulate user feedback, among other things. In the history of the world, no additional development has provided businesses with such an easy way of connecting many people while also providing products and services on a global scale at such a low cost.

4.4.4. Finances: It all comes down to translating the value that ITS providers deliver into revenues and profits. It includes the pricing techniques, how the company makes the best use of its tangible and intangible assets, and how the organisation converts the value of its goods and assets into monetary compensation. Increasingly, companies are placing a greater emphasis on intangible assets (such as their brand name, supplier network, intellectual property, and information value) as their primary source of value creation. In contrast, tangible, or physical assets account for a smaller proportion of overall firm value in the internet age. It, in turn, assists in cost reduction and the ability to operate more efficiently with the same or fewer resources available. Because of the nature of their company, a significant amount of money is required by several ITS companies. Even though establishing an information technology services company may require a substantial initial investment and may take a long time to recoup expenses and generate profits, an e-Business model that takes advantage of the convenience that the internet provides may accelerate the process of recouping costs and generating profits. In recent years, there have been countless examples of online businesses that have seen their revenues and corporate worth increase in a matter of years due to their successful usage of the internet. The previous three aspects influence and are influenced by the financial components. As a result, a long-term business plan must be developed to give enough returns and benefits to investors while also ensuring that donors are compensated. All business strategies must be self-financed after a certain point.

4.4.5. Partners credibility: ITS's long-term success is based on the support of all of its important people and groups. An essential part of stakeholder trust in ITS is how the possible benefits are spread out across different groups of people and how the benefits spread out across the whole society.

Consider the case of a small transportation provider who wants to deploy an electronic ticketing system (ETS). The cloud provider, the transportation operator, and the technology partner are the three primary actors in the system development method. SaaS, PaaS, and IaaS are the three levels of cloud services the Cloud Provider offers. The Microsoft Windows Azure platform is used for demonstration. The Technological Partner (TP) oversees constructing and maintaining the ticketing system and establishing a set of services for the transport operator to enable a customised business process. Because of the modular architecture, TP can use previously developed services to implement the required system to serve the business. The front-end device can also provide and use previously developed projects. The current concept is aimed toward a SaaS model, but it might be transformed into a PaaS or even a standalone solution with additional effort. The cloud provider's resources link two devices to the cloud (e.g., processors, message queues). The operator acquires front-end equipment (such as gates and validators) through the device provisioning approach, which TP registers on the cloud system. We're currently exploring a new process of integrating an Android device.

Each root is a multi-operator transport area where various aspects of the business information (clients, cards, etc.) are shared by many operators. The following are the rules that govern the tenancy hierarchy:

There is a lot of information about their clients available to people who live on the lower floors (operators). Upper-level tenants can read and consolidate common business information for the metropolitan area two or three times.

Third, upper-level renters aren't allowed to see private customer information, like customers' cards and purchases; and fourth, upper-level renters can't see information about lower-level renters (e.g., customers, cards, sales, and validations).

Tickets and validations can be sold and validated by people who are not at the event. There are many things that people can talk about when it comes to multi-tenancy.

Privacy, security, and the ability to add more features were the main factors [39]. It is essential to ensure that no one operator has access to information about other operators (they may be competitors). On the other hand, it is widespread for a business to need specific changes to its work.

When you use a different database design, you must figure out how to connect standard business information (like multi-modal ticket sales) on the upper levels, which is made at the lower levels, with the need to have a hierarchy of tenants. The master repository consist of the business information from lower levels. People at the bottom of the food chain don't get private information.

4.5   Market Trend of ITS

The market for intelligent transportation systems was worth USD 22.88 billion in 2020 and is predicted to grow to USD 30.65 billion by 2026, with a CAGR of 5.11% (2021–2026).

4.6   Societal Impact of ITS

Making our automobiles more "connected," individual privacy is undoubtedly a problem [39]. There is a slew of potential privacy issues arising from GPS tracking and connecting automobiles to each other and the internet, with a slew of societal ramifications. The influence of automobile automation on society has been the subject of much debate. Vehicle automation will likely open new mobility alternatives for the young, elderly, and crippled, in addition to the traditional worries of "will they be safe?" and "will traffic congestion be reduced?" It will undoubtedly improve our society's standard of living.

Cybersecurity is undeniably crucial for our computers and other electronic gadgets. A week passes without someone's latest server being hacked and private information being exposed. However, if we don't have the necessary cybersecurity safeguards, connected vehicles might suffer worse safety consequences. If someone maliciously broadcasted misleading fundamental safety messages, potentially dangerous crashes would almost certainly occur. As our ITS society grows, I believe we will have to cope with an increasing number of societal concerns associated with ITS technology. As a result, I'm taking steps to link our ITS society to the SSIT society and designate a representative from our organisation to participate actively in SSIT. Again, our job isn't only to do ITS research and enhance ITS technology; it's also to comprehend the policy and societal ramifications.

4.7   Conclusion

The situation in Russia differs from that in the United States. Before the road was created, the suburbs of most major cities had already been extensively built up with dwellings. In truth, new residential, commercial, and industrial development does not need a new road; instead, unique residential, commercial, and industrial development encourage the construction of a new road. What's more intriguing is that the construction of a new route, rather than the increased travel speeds, may have a more significant impact on land-use change in Russia. From this perspective, there is no reason to expect that the shift in ITS management of road charges will result in a land-use change.

User growth on toll roads and increased regional attractiveness may be confirmed after a certain amount of time. An investigation of these consequences could focus on our future research, as it will necessitate additional research and studies.

References

1. Lytras, Miltiadis D., Kwok T. Chui, and Ryan W. Liu. 2020. "Moving Towards Intelligent Transportation via Artificial Intelligence and Internet-of-Things" *Sensors* 20, no. 23: 6945. doi:https://doi.org/10.3390/s20236945
2. Zantalis, Fotios, Grigorios Koulouras, Sotiris Karabetsos, and Dionisis Kandris. 2019. "A Review of Machine Learning and IoT in Smart Transportation" *Future Internet* 11, no. 4: 94. doi:https://doi.org/10.3390/fi11040094
3. Victor Chang, An ethical framework for big data and smart cities, Technological Forecasting and Social Change, Volume 165, 2021, doi:https://doi.org/10.1016/j.techfore.2020.120559.
4. Leslie, David, Understanding Artificial Intelligence Ethics and Safety: A Guide for the Responsible Design and Implementation of AI Systems in the Public Sector (June 10,

2019). Available at SSRN: https://ssrn.com/abstract=3403301 or doi:https://doi.org/10.2139/ssrn.3403301

5. Golubchikov O, Thornbush M. Artificial Intelligence and Robotics in Smart City Strategies and Planned Smart Development. *Smart Cities*. 2020; 3(4):1133–1144. doi:https://doi.org/10.3390/smartcities3040056

6. Amador, O., Urueña, M., Calderon, M., Soto, I. Evaluation and improvement of ETSI ITS Contention-Based Forwarding (CBF) of warning messages in highway scenarios (2022) Vehicular Communications, 34, art. no. 100454. DOI: https://doi.org/10.1016/j.vehcom.2022.100454

7. Zhang, Y., Li, S., Blythe, P., Edwards, S., Guo, W., Ji, Y., Xing, J., Goodman, P., Hill, G. Attention Pedestrians Ahead: Evaluating User Acceptance and Perceptions of a Cooperative Intelligent Transportation System-Warning System for Pedestrians (2022) Sustainability (Switzerland), 14 (5), art. no. 2787. DOI: https://doi.org/10.3390/su14052787

8. Malik, A., Khan, M.Z., Faisal, M., Khan, F., Seo, J.-T. An Efficient Dynamic Solution for the Detection and Prevention of Black Hole Attack in VANETs (2022) Sensors (Basel, Switzerland), 22 (5). DOI: https://doi.org/10.3390/s22051897

9. Mfenjou, M.L., Ari, A.A.A., Njoya, A.N., Mbogne, D.J.F., Kolyang, Abdou, W., Spies, F. Control points deployment in an Intelligent Transportation System for monitoring inter-urban network roadway (2022) Journal of King Saud University – Computer and Information Sciences, 34 (2), pp. 16–26. DOI: https://doi.org/10.1016/j.jksuci.2019.10.005

10. Studer, L., Ponti, M., Agriesti, S., Sala, M., Gandini, P., Marchionni, G., Gugiatti, E. Modeling Analysis of Automated and Connected Cars in Signalized Intersections (2022) International Journal of Intelligent Transportation Systems Research, DOI: https://doi.org/10.1007/s13177-021-00279

11. Shen, S., Lv, C., Zhu, H., Sun, L., Wang, R. Potentials and Prospects of Bicycle Sharing System in Smart Cities: A Review (2022) IEEE Sensors Journal, DOI: https://doi.org/10.1109/JSEN.2022.3160178

12. Lakhan, A., Mohammed, M.A., Ibrahim, D A , Kadry, S., Abdulkareem, K.H. ITS Based on Deep Graph Convolutional Fraud Detection Network Blockchain-Enabled Fog-Cloud (2022) IEEE Transactions on Intelligent Transportation Systems, DOI: https://doi.org/10.1109/TITS.2022.3147852

13. Zhao, J., Chang, X., Feng, Y., Liu, C.H., Liu, N. Participant Selection for Federated Learning With Heterogeneous Data in Intelligent Transport System (2022) IEEE Transactions on Intelligent Transportation Systems, DOI: https://doi.org/10.1109/TITS.2022.3149753

14. Dogra, R., Rani, S., Babbar, H., Verma, S., Verma, K., Rodrigues, J.J.P.C. DCGCR: Dynamic Clustering Green Communication Routing for Intelligent Transportation Systems (2022) IEEE Transactions on Intelligent Transportation Systems, DOI: https://doi.org/10.1109/TITS.2022

15. Gao, Y., Ren, T., Zhao, X., Li, W. Sustainable Energy Management in Intelligent Transportation (2022) Journal of Interconnection Networks, DOI: https://doi.org/10.1142/S0219265921460099

16. Wang, J., Pradhan, M.R., Gunasekaran, N. Machine learning-based human-robot interaction in ITS (2022) Information Processing and Management, 59 DOI: https://doi.org/10.1016/j.ipm.2021.102750

17. Gaur, L., & Ramakrishnan, R. (2019). Developing internet of things maturity model (IoT-MM) for manufacturing. Int. J. Innovative Technol. Exploring Eng. (IJITEE), 9(1), 2473–2479.

18. Gaur Loveleen, Bhandari Mohan, Bhadwal Singh Shikhar, Jhanjhi Nz, Mohammad Shorfuzzaman, and Mehedi Masud. 2022. Explanation-driven HCI Model to Examine the Mini-Mental State for Alzheimer's Disease. ACM Trans. Multimedia Comput. Commun. Appl. (March 2022). doi:https://doi.org/10.1145/3527174

19. Ramakrishnan, R., & Gaur, L. (2019). Internet of things: approach and applicability in manufacturing. CRC Press.

20. Oberoi, S., Kumar, S., Sharma, R. K., & Gaur, L. (2022). Determinants of artificial intelligence systems and its impact on the performance of accounting firms doi:https://doi.org/10.1007/978-981-16-2354-7_38

21. Gaur, L., Afaq, A. Metamorphosis of CRM: Incorporation of social media to customer relationship management in the hospitality industry (2020) Handbook of Research on Engineering Innovations and Technology Management in Organizations, pp. 1–23. 9 DOI: https://doi.org/10.4018/978-1-7998-2772-6.ch001

22. Rana, J., Gaur, L., Singh, G., Awan, U. and Rasheed, M.I. (2021), "Reinforcing customer journey through artificial intelligence: a review and research agenda", International Journal of Emerging Markets, Vol. ahead-of-print No. ahead-of-print. doi:https://doi.org/10.1108/IJOEM-08-2021-1214

23. G. Singh, B. Kumar, L. Gaur and A. Tyagi (2019), "Comparison between Multinomial and Bernoulli Naïve Bayes for Text Classification," 2019 International Conference on Automation, Computational and Technology Management (ICACTM), pp. 593–596, doi: https://doi.org/10.1109/ICACTM.2019.8776800.

24. Gaur L., Agarwal V., Anshu K. (2020), "Fuzzy DEMATEL Approach to Identify the Factors Influencing Efficiency of Indian Retail," Strategic System Assurance and Business Analytics. Asset Analytics (Performance and Safety Management). Springer, Singapore. doi:https://doi.org/10.1007/978-981-15-3647-2_

25. Gaur, L., Afaq, A., Singh, G. and Dwivedi, YK (2021), "Role of artificial intelligence and robotics to foster the touchless travel during a pandemic: a review and research agenda", International Journal of Contemporary Hospitality Management, Vol. 33 No. 11, pp. 4079–4098. doi:https://doi.org/10.1108/IJCHM-11-2020-1246

26. Sharma, S., Singh, G., Gaur, L., & Sharma, R. (2022). Does psychological distance and religiosity influence fraudulent customer behaviour? International Journal of Consumer Studies, doi:https://doi.org/10.1111/ijcs.12773

27. Sahu, G., Gaur, L., & Singh, G. (2021). Applying niche and gratification theory approach to examine the users' indulgence towards over-the-top platforms and conventional TV. Telematics and Informatics, 65 doi:https://doi.org/10.1016/j.tele.2021.101713

28. Singh, G., Gaur, L., & Ramakrishnan, R. (2017). Internet of Things – technology adoption model in India. Pertanika Journal of Science & Technology, 25(3), 835–846.

29. Chaudhary, M., Gaur, L., Jhanji, N. Z., Masud, M., & Aljahdali, S. (2022). Envisaging employee churn using MCDM and machine learning. Intelligent Automation and Soft Computing, 33(2), 1009–1024. doi:https://doi.org/10.32604/iasc.2022.023417

30. Loveleen Gaur, Anam Afaq, Arun Solanki, Gurmeet Singh, Shavneet Sharma, N.Z. Jhanji, Hoang Thi My, Dac-Nhuong Le, Capitalising on big data and revolutionary 5G technology: Extracting and visualising ratings and reviews of global chain hotels, Computers & Electrical Engineering, Volume 95, 2021, 107374, ISSN 0045-7906, doi:https://doi.org/10.1016/j.compeleceng.2021.107374

31. Gaur L, Singh G, Solanki A, Jhanji NZ, Bhatia U, Sharma S, et al. Disposition of youth in predicting sustainable development goals using the neuro-fuzzy and random forest algorithms. Human Centric Computer Information Science. (2021) 11:24. DOI: https://doi.org/10.22967/HCIS.2021.11.024 ranks 25/162 (Q1)

32. R. Ramakrishnan and L. Gaur, "Smart electricity distribution in residential areas: Internet of Things (IoT) based advanced metering infrastructure and cloud analytics," 2016 International Conference on Internet of Things and Applications (IOTA), 2016, pp. 46–51, doi: https://doi.org/10.1109/IOTA.2016.7562693

33. Chaki, S., Ahmed, S., Easha, N.N., Biswas, M., Sharif, G.T.A. and Shila, D.A., 2021, August. A Framework for LED Signboard Recognition for the Autonomous Vehicle Management System. In 2021 International Conference on Science & Contemporary Technologies (ICSCT) (pp. 1–6). IEEE.

34. Biswas, M., Chaki, S., Ahammed, F., Anis, A., Ferdous, J., Siddika, A.M., Shila, D.A. and Gaur, L., 2022, February. Prototype Development of an Assistive Smart-Stick for the Visually

Challenged Persons. In 2022 2nd International Conference on Innovative Practices in Technology and Management (ICIPTM) (Vol. 2, pp. 477–482). IEEE.

35. Rahaman, M.N., Biswas, M.S., Chaki, S., Hossain, M.M., Ahmed, S. and Biswas, M., 2021, December. Lane Detection for Autonomous Vehicle Management: PHT Approach. In 2021 24th International Conference on Computer and Information Technology (ICCIT) (pp. 1–6). IEEE.

36. Ramakrishnan, Ravi, Loveleen Gaur, and Gurinder Singh. 2016. "Feasibility and Efficacy of BLE Beacon IoT Devices in Inventory Management at the Shop Floor." International Journal of Electrical and Computer Engineering (IJECE) 6 (5): 2362–2368. doi:https://doi.org/10.11591/ijece.v6i5.10807

37. Dinesh K. Sharma, Loveleen Gaur and Daniel Okunbor, 2007. "Image compression and feature extraction with neural network", Proceedings of the Academy of Information and Management Sciences, 11(1): 33–38

38. Ramakrishnan, R., Gaur, L. (2016). Application of Internet of Things (IoT) for Smart Process Manufacturing in Indian Packaging Industry. In: Satapathy, S., Mandal, J., Udgata, S., Bhateja, V. (eds) Information Systems Design and Intelligent Applications. Advances in Intelligent Systems and Computing, vol 435. Springer, New Delhi. doi:https://doi.org/10.1007/978-81-322-2757-1_34

39. Santosh, K., Gaur, L. Privacy, Security, and Ethical Issues (2021) Springe Briefs in Applied Sciences and Technology, pp. 65–74. DOI: https://doi.org/10.1007/978-981-16-6768-8_8

Chapter 5
Explainable AI in ITS: Ethical Concerns

5.1 AI and Sustainability

AI is the ally that sustainable development [1–3] requires inventing, executing, advising, and planning our planet's future and sustainability more effectively. AI will, among other things, help us build more efficiently, use resources more sustainably, and decrease and manage waste more effectively. Combining AI and sustainable development will assist all industries in building a better planet that meets current demands while avoiding putting future generations at risk owing to climate change or other significant concerns. AI is already used to make considerably [4] more dynamic and accurate estimates of a territory's infrastructure requirements. The National Infrastructure Systems Model (NISMOD) in the United Kingdom, for example, tackles numerous domains such as energy, digital, transportation, and water management. NISMOD is much better at dealing with interdependencies across different forms of infrastructure and enables real-time data refreshment. The sensing capability employs knowledge representation and deep learning [5] and uses AI [6].

AI assists in [8] cleaner transportation networks and anticipating extreme weather conditions. AI has the potential to speed up worldwide efforts to protect the environment and conserve [9] resources. AI applications could also help design more energy-efficient buildings, power storage improvements, and the best deployment of renewable energy by putting solar and wind electricity into the grid [10]. Simultaneously, AI levers might reduce global greenhouse gas (GHG) emissions by 4% in 2030. Deep learning [7], predictive skills, and intelligent grid systems can help AI manage renewable energy supply and demand. Finally, AI can aid in the reduction of traffic congestion, the improvement of cargo transportation, and develop autonomous (or self-driving) vehicles. AI is essential because it enables software to do human-like skills at a lower cost.

L. Gaur, B. M. Sahoo, *Explainable Artificial Intelligence for Intelligent Transportation Systems*, https://doi.org/10.1007/978-3-031-09644-0_5

Apart from the energy sector, AI has the potential to aid a wide range of industries and businesses while also benefiting the environment. For example, it is used in agriculture to improve irrigation and fertilisation efficiency, humidity, temperature, and fertilisation sensors. AI can help to forecast crop requirements shortly. Drones for monitoring and hyperspectral image analysis for comprehensive pest management are among the most modern agricultural sustainability options.

AI enables more efficient storage, manufacturing [8] and distribution methods for the sector. Furthermore, artificial vision systems in manufacturing [9] detect defects in assembly lines that are invisible to the naked eye and safety flaws or potential disasters. The latter is particularly critical in industries where safety is paramount, such as construction. The technology created in tunnel boring machines, highly complicated devices, is an example of applying AI to promote higher sustainability. A breakdown can halt all or a significant portion of underground construction [10]. The cost reductions and efficiency gained from large-scale tunnel drilling will be unprecedented. AI [11] is thus a critical component in predicting errors and minimising issues that could jeopardise long-term progress.

5.2 Transportation Network Applications

AI dynamically optimise transport network control to increase speed and capacity without any physical additions. Norfolk Southern Corporation in the United States is one of many examples, as it dynamically optimises traffic patterns across the rail freight network, allowing train drivers to boost speed by 10% to 20%. The Think feature, which uses knowledge reasoning, provides the processing of significantly larger volumes of real-time data than a human dispatcher could handle.

Changi Airport in Singapore features a prototype AI-powered system for predicting long-haul flight arrival times with far higher precision than possible, reaching 95% accuracy. ML [12, 40] and decision allow for more responsive and efficient ground operations and shorter passenger lines.

ITS is one area where AI applications have advanced quickly. These frameworks suggest that various innovations and communication frameworks are used to move forward in driving involvement. They collect critical data used by ML [13, 14] for forecasting. As it develops, data complexity will increase, and deep learning techniques will become increasingly crucial for discovering designs and features in this data to achieve a connected transportation framework.

5.2.1 Accidents Detection

Several attempts have characterised an accident's time, location, and severity to improve traffic management and reduce congestion. These methods range from manual reports to Neural Networks [15] algorithms that run automatically. Manual statements made by people can delay the discovery and cost-effectiveness of events. On the other hand, algorithms can gather data from sensors along the road to compute the flow of features before and after a collision. Statistical techniques like the California Algorithm incident detection are initially used to establish accident detection systems. However, using an algorithm on arterial roads is problematic due to street parking and traffic signals. As a result, neural network methods based on algorithms are developed. A neural network classification algorithm [16] is evaluated to detect accidents on a freeway.

5.2.2 AI Predictive Models

With the rapid evolution of ITS, new technologies such as IoT, ML, AI [4, 7, 8, 19, 28, 29] for anticipating traffic data have been developed. Modern traveller information systems, complicated traffic management systems, advanced public transportation systems, and commercial vehicle operations rely on this technology. Historical data acquired from sensors connected to the roads are used to build intelligent predictive systems [17]. Instead, these data feed machine learning and artificial intelligence algorithms for real-time, short-term, and long-term forecasting [18–20]. Previously, research concentrated on utilising a rudimentary feedforward neural network to predict short-term streams. Another study in Queensland, Australia, used a 1.5-kilometre stretch of highway data. A persistent time-lag network is used to create an object-oriented version of the neural network (TLRN). [With 90–94% accuracy, the program anticipated speed for the next 5 minutes].

5.2.3 AI in the Airline Industry

In the airline industry [21], AI is more successful at managing aeroplane journeys. AI may help with technology, such as machine learning, software/hardware, and numerous applications, such as intelligent maintenance and aviation route optimisation. PLADS is a system that collects data from extremely dense aircraft reports and modifies it to help the vector machine and SA algorithm [22] systems. It demonstrates that SVM is effective for this type of classification. In contrast, a study found that using an unsupervised ML [23] system when landing an aeroplane can improve safety.

5.2.4 AI in Security

Another investigation, on the other hand, used the Probabilistic Neural Network (PNN) [14, 16] to verify the aircraft's security by testing the engine onboard. PNN learning for the network necessitates using a radial base function rather than a linear model in the RBN's hidden layer. Furthermore, a study produced an automated supervised Random Forest system [3] for more accurate identification of aviation turbulence, which can assist the pilot in avoiding straying from the set path, lower fuel costs [24], and improve air control management.

5.2.5   AI in Automatic Vehicle Location

The tracking of automobiles on transportation networks is another area where AI [25] solutions have proven useful. An Automatic Vehicle Location (AVL) system can improve public transportation's operating efficiency, control, and overall service quality. This system can collect data in real-time by following transportation units using GPS signals, detecting difficulties, advising vehicles of changes, and managing alternate routes. AI has transformed the transportation business. AI assists automobiles, trains, ships, and planes and smoothing traffic patterns. Technology has the potential to make all modes of transportation safer, cleaner, brighter, and more efficient. For example, AI-assisted [26, 27] autonomous mobility may reduce the number of traffic collisions caused by [human mistakes]. However, these benefits accompany serious concerns regarding unintended consequences and misuses, such as cyber-attacks and skewed mobility decisions. There are also job ramifications and ethical questions concerning AI's responsibility for decisions made in the absence of humans.

5.3   Autonomous Driving Levels and Enablers

Apart from assisting with driving, the DAS attempts to improve the driver's comfort and sense of security. Sensor data and a personalisation [28] module is used for the vehicle's performance to the driver's preferences. Security, safety, and comfort are the three basic foundations of personalisation. Security attempts to safeguard the car from theft or other purposeful damage. At the same time, comfort aims to tailor the car characteristics (e.g., car seat and mirror position, interior, temperature, etc.) to the driver's preferences. There are five significant levels of automated driving described below (Fig. 5.1).

5.3.1   Level 0: There Is No Automation

The bulk of today's autos are Level 0: manually operated. Despite the presence of technology to aid the driver, the "dynamic driving task" is carried out by humans. Because it does not "steer" the car, the emergency braking system, for example, does not qualify as automation.

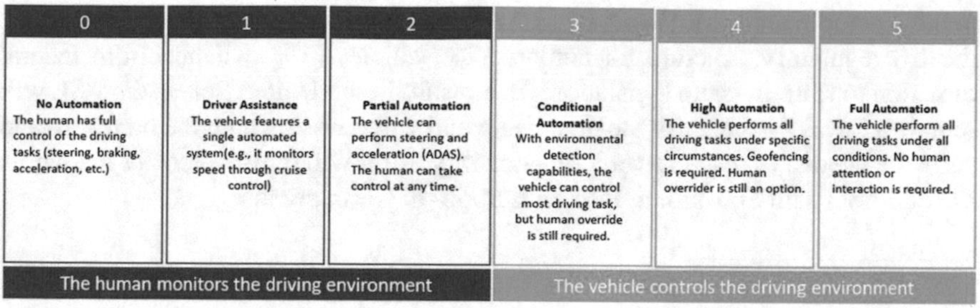

0	1	2	3	4	5
No Automation The human has full control of the driving tasks (steering, braking, acceleration, etc.)	**Driver Assistance** The vehicle features a single automated system(e.g., it monitors speed through cruise control)	**Partial Automation** The vehicle can perform steering and acceleration (ADAS). The human can take control at any time.	**Conditional Automation** With environmental detection capabilities, the vehicle can control most driving task, but human override is still required.	**High Automation** The vehicle performs all driving tasks under specific circumstances. Geofencing is required. Human overrider is still an option.	**Full Automation** The vehicle perform all driving tasks under all conditions. No human attention or interaction is required.

The human monitors the driving environment	The vehicle controls the driving environment

Fig. 5.1   5 levels of driving automation

5.3.2   Level 1: Driver Assistance

It is the simplest type of automation. The vehicle has a single automated driving aid system that includes steering and acceleration (cruise control). Adaptive cruise control, which keeps the car at a safe distance behind the next car [29], qualifies as Level 1 since the human driver monitors other aspects of driving, such as steering and braking.

5.3.3   Level 2: Partial Automation

ADAS stands for advanced driving assistance systems. Both steering and acceleration and deceleration are controlled by the vehicle. Because a human sits in the driver's seat and can take control of the car at any time, automation falls short of self-driving. Tesla Autopilot and Cadillac's Level 2 technology are examples (General Motors) Exceptional Cruise.

5.3.4   Level 3: Conditional Automation

The leap from Level 2 to Level 3 is significant from a technology standpoint, but it is negligible, if not non-existent, from a human perspective. Level 3 vehicles can monitor their surroundings and make intelligent [30] judgments, like accelerating past a slow-moving vehicle. They do, however, necessitate human intervention. If the machine fails to complete the task, the driver must remain attentive and ready to take charge. Audi (Volkswagen) announced nearly 2 years ago that the next generation of their flagship automobile, the A8, would be the first production Level 3 vehicle globally, and they delivered. This fall, commercial dealerships will be able to order the 2019 Audi A8L.

To minimise traffic congestion, "Traffic Jam Pilot" combines a lidar scanner with superior sensor fusion and computing power (plus built-in redundancies should make a component fail). However, as Audi was creating their engineering marvel, the US regulatory structure for autonomous vehicles [31] switched from federal direction to state-by-state legislation. As a result, in the United States, the A8L will remain a Level 2 vehicle for the time being and will come without the hardware and software needed to reach Level 3 functionality. Audi will, however, offer the whole Level 3 A8L with Traffic Jam Pilot in Europe (in Germany first).

5.3.5 Level 4: High Level of Automation

The main difference between Level 3 and Level 4 automation is that Level 4 automobiles may react if the system fails or something goes wrong. As a result, vehicles do not necessitate human engagement [32]. However, a human can still override the procedure manually, as and when needed. Self-driving cars are possible, but only in selected areas unless regulations and infrastructure are in place (usually an urban environment where top speeds reach an average of 30mph). Geofencing is the term for this. As a result, most of today's Level 4 vehicles are builts for ride-sharing.

5.3.6 Level 5: Full Automation

The "dynamic driving duty" is not incorporated in Level 5 automation. Hence human supervision is not needed. Level 5 vehicles will be devoid of steering wheels and pedals for acceleration and braking [33, 34]. Geofencing will be lifted, giving them the freedom to go wherever they want and do whatever an excellent human driver can. Fully driverless vehicles are tested worldwide but are not available for sale. The Society of Automotive Engineers (SAE) has provided six driving automation levels, ranging from zero to five (fully autonomous).

5.4 Personalised Mobility Services and AI

Although AI has roots dating back to the 1950s, tremendous progress has been made in the last 10 years due to greater processing power and data availability. The two features are the most significant enablers for AI algorithms to control innovative services and automate tough jobs. Almost every primary cloud provider has invested in AI to create a new market for AI solutions as a service, which can be used in various industries. To attract new clients and allow third-party organisations to build AI solutions, IBM introduced Watsonas a service. IBM Watson began as a question-and-answer system that could interpret natural language and react with information from a vast knowledge store and has now grown into a suite of enterprise-ready AI services, applications, and tools. The IBM Watson enables businesses to accelerate research and discovery, enrich interactions, anticipate and prevent disruptions, confidently recommend, scale expertise and learning, detect liabilities, and mitigate risk, allowing employees to focus on high-value work critical to an organisation's success. The Watson suite includes question-answering systems that extract facts from company documents, virtual assistants that reply to online customers, chatbots, and clever readers of complex documents (such as contracts).

Einstein AI is Salesforce's intelligent CRM [32] solution. The platform employs ML and AI to help managers make better decisions, increase employee productivity, and make personalised recommendations that boost consumer happiness. Companies can build solutions that discover significant patterns and trends in sales data. It helps to understand their customers by learning which channels, messages, and content they prefer; and allows every employee to have instant access to intelligent insights and business AI-powered analytics [35] using a combination of AI tools available as a service (Fig. 5.2).

Nvidia (Computer systems design service firm) has invested in hardware acceleration of deep learning architectures for autonomous vehicles and driver assistance since 2015. The Nvidia Drive AGX open independent vehicle computing platform gathers data from cameras, lidars, ultrasonic sensors, and radars, among other sources. The data is then processed to obtain a real-time 360-degree understanding of the surrounding environment, detect the vehicle's location on the map and within the surrounding area, and determine the next safe maneuver. This energy-efficient, high-performance computing architecture can create safe and responsive self-driving models. Data availability and connectivity will enable real-time sensing and traffic prediction applications, promoting peer-to-peer ride-sharing and self-driving automotive applications, according to Stanford's One Hundred Year Study for AI and Life in 2030. Also, people's trust in the safety and robustness of physical hardware drives the adoption of autonomous transportation, which predicts to grow from automobiles and trucks to flying vehicles and personal robots by 2030. According to the same study, the growth of personalised mobility services based on peer-to-peer interaction and car-sharing will reduce the necessity for car ownership and turn the car from an asset into service. As a result, the "vehicle as a service" concept will

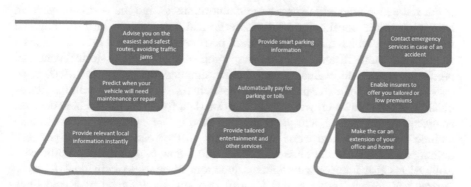

Fig. 5.2 Personalised mobility services

evolve. Companies that provide large-scale ride-sharing services, such as Uber and Lyft, have also used AI services. Uber's pricing optimisation system leverages historical data and location and trip data from its network of drivers and clients to forecast rider demand and ensure that sharp price changes used to limit rider demand are no longer necessary. Uber's ML models can also anticipate estimated UberEATS meal delivery times and the best pick-up locations and detect fraud. On-demand transportation firms use AI algorithms to connect drivers and passengers based on their site and reputation.

They also use pricing algorithms to change ride prices dynamically to maintain demand and supply equilibrium. Carpooling and ride-sharing services like Zimride and Nuride allow passengers to share routes and costs, but they've only gained a small amount of traction because they don't yet interface with public transportation. A combination of adaptive public transit and personal or shared rides can provide personalised mobility. It would be possible to deliver accurate real-time information for the public and shared transportation to passengers using data analytics and predictive algorithms. When used in conjunction with empty autonomous vehicles, it would be possible to cut passenger wait times and provide individualised transportation services from door to door. Shanghai Bus is a public transportation project established by the Shanghai Municipal Transport Commission that uses an Automatic Vehicle [35] Location system to offer passengers real-time information on the whereabouts of all public vehicles in the bus, train, and metro networks.

Entirely autonomous vehicles are still a few years away from mass manufacture, and their use is restricted to highways with a more controlled driving environment. The California Department of Motor Vehicles, on the other hand, established laws in April 2018 that allow for public testing and deployment of autonomous driverless automobiles. Self-driving trucks appear to be a more feasible undertaking, owing to their ability to operate in enclosed areas (such as a factory) or execute a repeated duty. Finally, many European countries are seeing the introduction of driverless buses (typically with an extra driver for safety reasons). They usually follow predetermined routes, but they employ sensors, cameras, GPS, and AI models to analyse visual input to detect and avoid impediments and securely transport passengers to their destinations. One of the first AI applications in transportation networks is intelligent route planning, which uses several algorithms to find the best path between two points while considering network traffic and performance [36]. Multimodal route planning, which mixes public transit, personal vehicles, and ride-sharing, is a significantly more difficult task requiring complex algorithms, historical data, and prediction models to evaluate many factors simultaneously. Google Maps is an excellent example of an AI-powered tool that helps drivers get to their destination as quickly as possible by crowdsourcing location data from Google-enabled devices, analysing vehicle speed, and providing real-time traffic estimates. They can also include reports on construction and incidents on the road. Another AI-powered option projected to decentralise and reduce traffic flow is personalised traffic control systems that interact with lights at junctions to adapt to individual vehicles. Surtrac (Intelligent Traffic Signal Contro) is a system designed by Rapid Flow Technologies for urban areas to respond to changing traffic in real-time, optimising traffic flows every second. Compared to typical traffic signal timing, Surtrac cuts travel times by 25%, waits at signals by 40%, stops by 30%, and emissions by 20%. It also optimises for pedestrians, bikes, public transportation, and connected cars, resulting in fewer waits for everyone.

The system design includes sensors at the intersection, a scheduler that assigns green time to the incoming car or pedestrian flows, and an executor that puts the selected schedules into action by connecting with traffic lights at the intersection and surrounding intersections [37]. Onboard diagnostics (OBD) systems have improved a vehicle's self-diagnostic and reporting capabilities, becoming a standard feature in cars. Although the original goal of OBD sensors was to monitor the vehicle's health and aid in the early detection of faults or dangers, they have lately been expanded with more sensing devices that collect data for both the car and the driver's state and behaviour.

Biosensors, wearable gadgets, and onboard cameras are used in personalised driver assistance systems (DAS), which offer real-time data on the driver's vital signs. Early detection of sleepiness, blackout, level of weariness, attention, or risk is possible because of a mix of data mining algorithms trained on several drivers' historical data and real-time information from each drive. For example, a camera can capture the driver's face every few milliseconds, then extract various picture features (along with any other biometric features like heart rate, blood pressure, and so on) and feed them to a pre-trained classification algorithm. The model will tell if the driver is awake or on the verge of falling asleep [38]. In a somewhat different example, the driver's reflexes on the road (e.g., braking speed or wheel-steering) can be recorded with car embedded sensors and fed into a model that determines whether the driver is weary and needs a break or not. If the response is not time-critical, the data is transferred to a cloud processing node (cloud computing paradigm); however, the processing must happen in the car in time-critical circumstances. With powerful GPU-enabled processors embedded in vehicles or the network provider's backbone, this is now possible (edge and fog computing models)[39].

5.5 Conclusion

The application of ML and AI algorithms on data collected by multiple sensors attached to drivers, cars, and the road network allows new intelligent solutions to be built. It also attracts extensive software development, sensor, and computer hardware businesses that push solutions as a service and works with car manufacturers to provide a variety of driver aid accessories. On the other hand, as new intelligent mobility services emerge and AI apps take over as drivers or co-pilots, many new concerns occur. Because the backbone of the connected autonomous vehicles network becomes a target for hackers, network providers must explore data encryption and anonymisation technologies to protect drivers' sensitive data while increasing network safety and security. Ride-sharing firms' growth also presents policy and legal challenges, such as competition with taxis and the necessity for a regulatory framework.

Autonomous driving systems replace driver-assistance systems. Several ethical questions arise, such as how autonomous cars will act in critical decision-making situations involving humans, such as when they must choose between actions that will cause human injury or death. Public transportation is likely to be replaced or supplemented by shared transportation and personal rapid transit. The growth of customised solutions will result in a massive increase in unused transportation capacity, which must be avoided. The multi-objective optimisation methods must be devised to deal with the redistribution of empty vehicles and limit the utilisation of empty cars.

References

1. https://www.europarl.europa.eu/RegData/etudes/BRIE/2019/635609/EPRS_BRI(2019)635609_EN.pdf
2. Lytras, Miltiadis D., Kwok T. Chui, and Ryan W. Liu. (2020), "Moving Towards Intelligent Transportation via Artificial Intelligence and Internet-of-Things" *Sensors* 20, no. 23: 6945. doi:https://doi.org/10.3390/s20236945
3. Gaur L, Singh G, Solanki A, Jhanjhi NZ, Bhatia U, Sharma S, et al. Disposition of youth in predicting sustainable development goals using the neuro-fuzzy and random forest algorithms. Human Centric Computer Information Science. (2021) 11:24. DOI: https://doi.org/10.22967/HCIS.2021.11.024 ranks 25/162 (Q1)
4. Zantalis F, Grigorios K, Sotiris K, and Dionisis K. (2019), "A Review of Machine Learning and IoT in Smart Transportation" *Future Internet* 11, no. 4: 94. doi:https://doi.org/10.3390/fi11040094
5. Gaur, L., Afaq, A., Singh, G. and Dwivedi, YK (2021), "Role of artificial intelligence and robotics to foster the touchless travel during a pandemic: a review and research agenda", International Journal of Contemporary Hospitality Management, Vol. 33 No. 11, pp. 4079–4098. doi:https://doi.org/10.1108/IJCHM-11-2020-1246
6. Noor Zaman, Loveleen Gaur, and Mamoona Humayun, "Approaches and Applications of Deep Learning in Virtual Medical Care", with IGI Global USA 2021, Release October 2021. (https://www.igi-global.com/book/approaches-applications-deep-learning-virtual/274538) ISBN13: 9781799889298|ISBN10: 1799889297|EISBN13: 9781799889304
7. Ramakrishnan, R., Gaur, L. (2016). Application of Internet of Things (IoT) for Smart Process Manufacturing in Indian Packaging Industry. In: Satapathy, S., Mandal, J., Udgata, S., Bhateja, V. (eds) Information Systems Design and Intelligent Applications. Advances in Intelligent Systems and Computing, vol 435. Springer, New Delhi. doi:https://doi.org/10.1007/978-81-322-2757-1_34
8. Gaur, L., & Ramakrishnan, R. (2019). developing internet of things maturity model (IoT-MM) for manufacturing. Int. J. Innovative Technol. Exploring Eng.(IJITEE), 9(1), 2473–2479.
9. Victor C, "An ethical framework for big data and smart cities", *Technological Forecasting and Social Change,* Volume 165, 2021, doi:https://doi.org/10.1016/j.techfore.2020.120559.
10. Rana, J., Gaur, L., Singh, G., Awan, U. and Rasheed, M.I. (2021), "Reinforcing customer journey through artificial intelligence: a review and research agenda", International Journal of Emerging Markets. doi:https://doi.org/10.1108/IJOEM-08-2021-1214
11. Chaudhary M, Gaur L, Jhanjhi NZ, Masud M, Aljahdali S, Envisaging Employee Churn Using MCDM and Machine Learning, Intelligent Automation and Soft Computing, Volume 33, 2021, 101713, ISSN 1079-8587, DOI:https://doi.org/10.32604/iasc.2022.023417

12. Leslie D, "Understanding Artificial Intelligence Ethics and Safety: A Guide for the Responsible Design and Implementation of AI Systems in the Public Sector" (June 10, 2019). Available at SSRN: https://ssrn.com/abstract=3403301 or doi:https://doi.org/10.2139/ssrn.3403301

13. K C Santosh, Loveleen Gaur, Artificial Intelligence and Machine Learning in Public Healthcare: Opportunities and Societal Impact ISBN 978-981-16-6767-1 https://www.springer.com/gp/book/9789811667671

14. Gaur, L., Bhatia, U., Jhanjhi, N.Z. et al., Medical image-based detection of COVID-19 using Deep Convolution Neural Networks. Multimedia Systems (2021). doi:https://doi.org/10.1007/s00530-021-00794-6

15. Golubchikov O, Thornbush M. "Artificial Intelligence and Robotics in Smart City Strategies and Planned Smart Development". *Smart Cities*. 2020; 3(4):1133–1144. doi:https://doi.org/10.3390/smartcities3040056

16. Dinesh K. Sharma, Loveleen Gaur and Daniel Okunbor, 2007. "Image compression and feature extraction with neural network", Proceedings of the Academy of Information and Management Sciences, 11(1): 33–38

17. Anam Afaq, Loveleen Gaur, The Rise of Robots to Help Combat Covid-19 (2021) 2021 International Conference on Technological Advancements and Innovations (ICTAI), pp. 69–74, IEEE

18. Sharma, S., Singh, G., Gaur, L., & Sharma, R. (2022). Does psychological distance and religiosity influence fraudulent customer behaviour? International Journal of Consumer Studies, doi:https://doi.org/10.1111/ijcs.12773

19. Gaur, L., Singh, G. and R. Ramakrishnan (2017). Bidirectional elearning using IoT smart mirrors in understanding consumer preference. Pertanika Journal of Science & Technology, 25 (3): 939– 948.

20. Qureshi, K.N. and Abdullah, A.H., 2013. A survey on intelligent transportation systems. *Middle-East Journal of Scientific Research*, 15(5), pp. 629–642.

21. Afaq, A., Gaur, L., Singh, G., & Dhir, A. (2021). COVID-19: Transforming air passengers' behaviour and reshaping their expectations towards the airline industry. Tourism Recreation Research, doi:https://doi.org/10.1080/02508281.2021.2008211

22. Ramakrishnan, R., Gaur, L., & Singh, G. (2016). Feasibility and efficacy of BLE beacon IoT devices in inventory management at the shop floor. International Journal of Electrical and Computer Engineering, 6(5), 2362–2368. doi:https://doi.org/10.11591/ijece.v6i5.10807

23. Gaur L, Bhandari M, Razdan T, Mallik S and Zhao Z (2022) Explanation-Driven Deep Learning Model for Prediction of Brain Tumour Status Using MRI Image Data. Front. Genet. 13:822666. doi: https://doi.org/10.3389/fgene.2022.822666

24. Neelakandan, S., Prakash, M., Bhargava, S., Mohan, K., Robert, N.R. and Upadhye, S., 2022. Optimal Stacked Sparse Autoencoder Based Traffic Flow Prediction in Intelligent Transportation Systems. In Virtual and Augmented Reality for Automobile Industry: Innovation Vision and Applications (pp. 111–127). Springer, Cham.

25. Loveleen Gaur, Anam Afaq, Arun Solanki, Gurmeet Singh, Shavneet Sharma, Nz Jhanjhi, Hoang Thi My, Dac-Nhuong Le, Capitalising on big data and revolutionary 5G technology: Extracting and visualising ratings and reviews of global chain hotels, Computers & Electrical Engineering, Volume 95, 2021, 107374, ISSN 0045-7906, doi:https://doi.org/10.1016/j.compeleceng.2021.107374

26. Singh, G., Gaur, L. and R. Ramakrishnan (2017). Internet of Things – technology adoption model in India, Pertanika Journal of Science & Technology, 25(3):835–846

27. Alvarez-Mamani, E., Vera-Olivera, H. and Soncco-Álvarez, J.L., 2022. Clustering Analysis for Traffic Jam Detection for Intelligent Transportation System. In *Annual International Conference on Information Management and Big Data* (pp. 64–75). Springer, Cham.

28. Gaur, L. (2019), Developing Internet of Things Maturity Model (IoT-MM) for Manufacturing, 'International Journal of Innovative Technology and Exploring Engineering (IJITEE)', ISSN: 2278–3075 (Online), Volume-9 Issue-1, November 2019, Page No. 2473–2479.

29. R. Ramakrishnan and L. Gaur, "Smart electricity distribution in residential areas: Internet of Things (IoT) based advanced metering infrastructure and cloud analytics," *2016 International Conference on Internet of Things and Applications (IOTA)*, 2016, pp. 46–51, doi: https://doi.org/10.1109/IOTA.2016.7562693

30. Gao, Y., Ren, T., Zhao, X. and Li, W., 2022. Sustainable Energy Management in Intelligent Transportation. *Journal of Interconnection Networks*, p. 2146009.

31. Singh, G., Gaur, L., & Agarwal, M. (2017). Factors Influencing the Digital Business Strategy. Pertanika Journal of Social Sciences and Humanities, 25(3), 1489–1500.

32. Gaur, L., Afaq, A. Metamorphosis of CRM: Incorporation of social media to customer relationship management in the hospitality industry (2020) Handbook of Research on Engineering Innovations and Technology Management in Organizations, pp. 1–23. 9 DOI: https://doi.org/10.4018/978-1-7998-2772-6.ch001

33. Santosh, K., Gaur, L. AI in Sustainable Public Healthcare (2021) Springer Briefs in Applied Sciences and Technology, pp. 33–40. DOI: https://doi.org/10.1007/978-981-16-6768-8_4

34. Santosh, K., Gaur, L. Privacy, Security, and Ethical Issues (2021) Springe Briefs in Applied Sciences and Technology, pp. 65–74. DOI: https://doi.org/10.1007/978-981-16-6768-8_8

35. Mohini Jain, Gurinder Singh, Loveleen Gaur, Green Internet of Things: Next-Generation Intelligence for Sustainable Development (2021), Advances in Interdisciplinary Research in Engineering and Business Management

36. Ravi Ramakrishnan and Loveleen Gaur, Innovation in Product Design: IoT Objects Driven New Product Innovation and Prototyping Using 3D Printers, Source Title: Additive Manufacturing: Breakthroughs in Research: 2020 |Pages: 21, DOI: https://doi.org/10.4018/978-1-5225-9624-0.ch014

37. G. Singh, B. Kumar, L. Gaur and A. Tyagi (2019), "Comparison between Multinomial and Bernoulli Naïve Bayes for Text Classification," 2019 International Conference on Automation, Computational and Technology Management (ICACTM), pp. 593–596, doi: https://doi.org/10.1109/ICACTM.2019.8776800.

38. Singh, G., Jain Vishal, Chatterjee, Jyotirmoy, Gaur, L. (2021), "Cloud and IoT Based Vehicular Ad-Hoc Networks", ISBN 1119761832, 9781119761839 https://books.google.co.in/books/about/Cloud_and_IoT_Based_Vehicular_Ad_Hoc_Net.html?id=XMCVzQEACAAJ&redir_esc=y

39. Singh, G., Gaur, L., & Ramakrishnan, R. (2017). Internet of Things – technology adoption model in India. Pertanika Journal of Science & Technology, 25(3), 835–846.

40. Chaudhary, M., Gaur, L., Jhanjhi, N. Z., Masud, M., & Aljahdali, S. (2022). Envisaging employee churn using MCDM and machine learning. Intelligent Automation and Soft Computing, 33(2), 1009–1024. doi:https://doi.org/10.32604/iasc.2022.023417

Printed in the United States
by Baker & Taylor Publisher Services